Forgotten Statistics:
A Self-Teaching Refresher Course

Douglas Downing, Ph.D.
School of Business and Economics
Seattle Pacific University
Seattle, Washington

Jeffrey Clark, Ph.D.
Mathematics Department Chair
Elon College
Elon College, North Carolina

BARRON'S

Barron's Educational Series, Inc.

All inquiries should be addressed to:
Barron's Educational Series, Inc.
250 Wireless Boulevard
Hauppauge, New York 11788

Library of Congress Catalog Card No. 96-21768
International Standard Book No. 0-8120-9713-0

Library of Congress Cataloging-in-Publication Data

Downing, Douglas.
    Forgotten statistics / by Douglas Downing and Jeffrey Clark.
        p.   cm.
    Includes index.
    ISBN 0-8120-9713-0
    1. Mathematical statistics.   I. Clark, Jeff.   II. Title.
QA276.12.D677   1996
519.5—dc20   96-21768
                                  CIP

PRINTED IN THE UNITED STATES OF AMERICA

987654321

Contents

Preface

An understanding of statistics is vital in several different fields (see Unit 1 for some examples). This book can help you in two ways: by giving you a narrative introduction to the essential topics in statistics (Part I), and by giving you an alphabetical reference section (Part II) where you can look up topics as you need them. In Part I, we act as if the subject of statistics is new to you. Even if you have seen statistics before, we assume that you decided to read this book because it has been a while since you studied it and would like a review.

Several Greek letters and other special symbols are used in statistics. You should become familiar with the symbols that are listed on page v so you will recognize them when they appear in statistical formulas.

Appendix A gives the most commonly used statistical tables (for the normal, chi-square, t, and F distributions). Appendix B covers the process of using a calculator, a computer statistics program, or a spreadsheet program to perform statistical calculations. In the old days statistics developed a reputation for being very difficult, which certainly was true when laborious calculations had to be done by hand. Now, however, you can use computational tools to free you from the drudgery, allowing you to concentrate on the concepts.

List of Symbols

Σ	(capital sigma): summation notation
μ	(mu): population mean
σ	(sigma): population standard deviation
σ^2	population variance
$\bar{x}$	sample mean
s	sample standard deviation
s^2	sample variance
$\Pr(A)$	probability that the event A occurs
$\Pr(A|B)$	conditional probability that event A occurs given that event B occurs
$!$	factorial (for example, $4! = 4 \times 3 \times 2 \times 1$)
$\binom{n}{j}$	number of combinations of j objects chosen from a group of n objects
$f(a)$	for a discrete random variable X, the probability function: $f(a) = \Pr(X = a)$
$f(x)$	for a continuous random variable X, the probability density function
$E(X)$	expected value of random variable X
$\text{Var}(X)$	variance of random variable X
$\text{Cov}(X, Y)$	covariance of X and Y
r^2	r-squared value: measuring how well a regression equation fits the data
r	correlation coefficient
π	(pi); approximately equal to 3.14159
e	approximately equal to 2.71828
$\sqrt{n}$	square root of n

PART I

Essential Concepts in
Statistics

UNIT 1

Introduction to Statistics

Statistics is a valuable tool to help you analyze data. This book assumes you have studied statistics sometime in the past and would like a review, or perhaps you are new to the subject and would like a book that both gives a concise introduction and some reference material.

There are two main parts of the book. The first eight units consist of a narrative introduction to key ideas in statistics. These units avoid bogging you down in all of the details, but they have enough information for you to see the highlights of the crucial concepts. The second part is an alphabetical reference section that you can turn to for more information about specific topics as needed.

Suppose you are working as a researcher studying human behavior. You will need observations of many people for the variables you are interested in: perhaps age, height, food products consumed, time spent in various activities, and so on. One problem you would face after collecting the data is that it is hard for a person to see the meaning in a long list of numbers. It helps if we can illustrate the numbers with a graph, and it helps to summarize the numbers (for example, calculate the average). The subject of *descriptive statistics* (covered in Unit 2) considers ways of summarizing data.

Units 3 to 8 cover *inferential statistics* (or *statistical inference*), where the problem is even harder: you usually are unable to obtain data about all of the items you are interested in. The complete set of items you are interested in is called the *population*. Typically it is too hard or expensive to investigate the entire population. Instead, you will have to content yourself with investigating a few items chosen from the population in a *sample*. In order for this to work, we must have some assurance that the sample will be representative of the population. This can be difficult, because any system we can think of for choosing the sample runs the risk of biasing it and making it unrepresentative. There are clearly some sample selection procedures that are not good. For example, we should not simply select people from our own home town, since they might be unrepresentative of people in the rest of the country. We would not know how unrepresentative in the absence of data about people in the rest of the country.

It turns out that the best sample selection system is to have no systematic approach at all —instead, select the sample totally at random. The ideal way would be to put the name of everyone in the population in a little capsule, put all of the capsules in a giant drum, mix them thoroughly, and then start selecting them at random. It is seldom possible in practice to follow this exact procedure, but the same concept applies to samples selected with computer-generated random number lists.

Your first reaction might well be that there is no guarantee that a randomly selected sample will be representative of the population. For example, it is possible that all of the people in your sample will be from your home town. However, you should realize that is unlikely. If the sample contains 1,000 people (a typical figure), chosen randomly from a population of over 200 million people, the chance that all 1,000 people will be from your home town is very, very small. In fact, the chance of the sample being unrepresentative in any manner is small, so we can have faith that it will give us an accurate picture of the population.

Statistical analysis is also used in the testing of a new medicine. It would not be a good idea simply to give the medicine to some people and see how they react, since that would not allow you to tell whether the medicine made a difference. Many people would recover from the disease anyway, with or without the medicine. In order to verify that the medicine is effective, you need to be able to show that the people with the medicine do better than the people without it. This requires a test conducted on two groups: a *treatment group* that receives the medicine, and a *control group* that does not. One risk you face with this kind of test is that the people in the treatment group might all be sicker than the control group; in that case, the people receiving the medicine might do worse even if the medicine really helps. The best way to avoid that problem is with random selection. Ideally, you would flip a coin to determine whether each individual will be in the control group or the treatment group. If this is done, there is still a possibility that the treatment group might be systematically different from the control group, but this is not very likely if the two groups are large enough.

You do need to be careful to avoid contaminating the results by psychological factors. If people knew which group they were in, then the treatment group might get better because they are psychologically disposed to think they are being helped, whereas the control group might give up in despair if they have the knowledge they are not being helped. The solution is to prevent the people from knowing which group they are in. This means that control group members are given an ineffective sugar pill (placebo) that is indistinguishable from the medicine the treatment group receives. Ideally, the doctors administering the medicines don't know which group is which, so they don't introduce psychological contamination. The survey is then said to be *double-blind*. (The researchers, of course, need careful records to know who is in the treatment group and the control group so they can compare the results.)

One might wonder about the ethics of involving the control group in the study when they receive nothing that might help them. The answer to this concern is that if you were sure that the medicine would help, you would not need to do the study. The study is required precisely because you don't know whether it will work—and one could also be concerned about the ethics of giving the untested medicine to the treatment group. The people who volunteer to be in such a study are providing valuable benefits to all future people who might benefit from that medicine.

Some other examples of statistical analysis include the following:

- The Census Bureau conducts a giant monthly survey (the Current Population Survey) that is used to determine the monthly unemployment rate and other facts about the population. The bureau also takes surveys of businesses so it can estimate the gross domestic product and other measures of the state of the economy.

- The ratings for television and radio stations are determined by samples. The ratings services select a random sample of households to measure viewership and listenership.

- Businesses will take surveys to learn about consumer preferences, and they will test market new products.

- Businesses will use statistical methods to conduct quality control checks of their products.

- Auditors will take random samples of documents to check when they are investigating the integrity of a company's books.

- Researchers in the social and natural sciences use statistical analysis to test relationships that are proposed in hypothetical models.

We need to develop precisely the concept of the accuracy of a random sample. To do that, we must first study probability theory (covered in Unit 3) and a specific probability concept known as a *random variable* (Unit 4). Unit 5 applies these ideas to the specific question of an opinion poll trying to predict an election, giving a formula that allows you to predict how accurate a poll is likely to be. Unit 6 covers the use of a sample to estimate a *confidence interval* for the unknown population mean. It also shows how to test a hypothesis about specific values of the mean.

Another important area of statistics is the use of *regression analysis* to test for the presence of a relationship between two or more variables. For example, we might like to know if there is a relationship between height and weight, or between income and spending. This topic is covered in Unit 7.

Finally, Unit 8 gives some warnings of areas where you can misuse statistics if you are not careful.

UNIT 2

Descriptive Statistics

MEAN AND MEDIAN

Descriptive statistics involves finding meaningful ways to summarize a collection of data. For example, suppose you measure your commuting time (in minutes) for five days with these results:

$$25, 38, 28, 26, 30$$

One way to summarize the data is to calculate the average:

$$\text{average} = \frac{25 + 38 + 28 + 26 + 30}{5} = \frac{147}{5} = 29.4$$

The average is also called the *mean*, symbolized by the Greek letter mu (μ), or by putting a bar over the quantity. It is found by adding up all elements and then dividing by the number of elements. If there are n numbers in the list (symbolized by $x_1, x_2, \ldots x_n$) the average is:

$$\mu = \bar{x} = \frac{x_1 + x_2 + \ldots + x_n}{n}$$

(For now we will use μ and $\bar{x}$ interchangeably to represent the average; later we will see that it is common to use μ to represent the mean of a population and $\bar{x}$ for the mean of a sample.)

The sum of x_1 to x_n can also be written with summation notation:

$$x_1 + x_2 + \ldots + x_n = \sum_{i=1}^{n} x_i$$

The symbol Σ, the Greek capital letter sigma, stands for summation; the designation "$i = 1$" at the bottom tells to start adding where i equals 1; and the "n" at the top tells to stop adding where i equals n. The expression written to the right of the sigma (x_i in this case) tells you what to add.

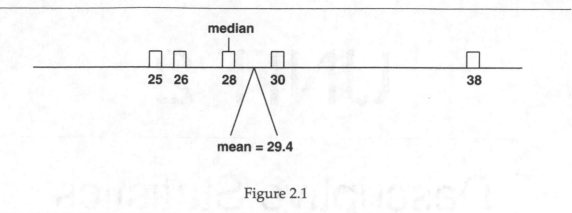

Figure 2.1

If we have a teeter-totter representing a number line, and we put a block at the location of each of the 5 numbers in our list, then the mean is the point where the teeter-totter will balance (see Figure 2.1).

Another way of representing the values of a list is to determine the *median,* or middle value of the list. If we put our five numbers in order, we see that 28 is the median value (see Figure 2.1). In this case there is an odd number of items in the list, so it is possible to determine a unique middle element. If there is an even number of items in the list, then the median will be the number halfway between the two middle numbers. For example, the median of the list $\{10, 14, 18, 19\}$ is $\frac{(14+18)}{2} = 16$.

You might wonder: Which is a better representation of the list, the mean or the median? If the numbers are symmetric about their middle, the mean and the median coincide, so it doesn't matter which you use. With our commuting data, note that one value, 38, is unusually large. From Figure 2.1 we see that the mean is larger than the median. This will happen when an unusually large value pulls the mean up. The unusually large value does not affect the median; for example, if the largest value in the list were 1,000,000, instead of 38, the median would still be 28. On the other hand, an unusually small value will pull the mean down below the median.

FREQUENCY HISTOGRAMS

Next we will investigate the age at inauguration of the presidents of the United States:

$$
\begin{array}{ccccccc}
57 & 61 & 57 & 57 & 58 & 57 & 61 \\
54 & 68 & 51 & 49 & 64 & 50 & 48 \\
65 & 52 & 56 & 46 & 54 & 49 & 50 \\
47 & 55 & 55 & 54 & 42 & 51 & 56 \\
55 & 51 & 54 & 51 & 60 & 62 & 43 \\
55 & 56 & 61 & 52 & 69 & 64 & 46 \\
\end{array}
$$

Our minds are unable to absorb much meaning from a long list of numbers like this. We can better understand the information if we draw a picture similar to the teeter-totter diagram

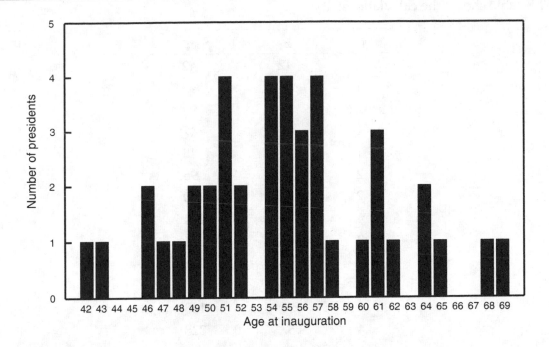

Figure 2.2

earlier, only now we recognize that some numbers occur more than once in the list. Therefore, for each age we will draw a bar whose height represents the number of presidents of that age (see Figure 2.2).

This type of diagram is called a *frequency histogram* or *bar diagram*. We will often draw this type of diagram to illustrate the nature of the distribution of a group of numbers. We can see from the diagram that ages near the middle (in the 50s) are more common than the extreme ages (low 40s and high 60s).

If we let f_i be the frequency of x_i (that is, the number of times that the age x_i occurs in the list), then we can find the mean by the following formula:

$$\overline{x} = \frac{\displaystyle\sum_{i=1}^{m} f_i x_i}{\displaystyle\sum_{i=1}^{m} f_i}$$

The letter m represents the number of different ages represented in the list. If we again let n represent the total number of items in the list (in this case, $n = 42$ because there are 42 presidents), then:

$$n = \sum_{i=1}^{m} f_i$$

So, by substitution:

$$\overline{x} = \frac{\displaystyle\sum_{i=1}^{m} f_i x_i}{n}$$

This table shows the calculations:

f_i	x_i	$f_i x_i$
1	42	42
1	43	43
2	46	92
1	47	47
1	48	48
2	49	98
2	50	100
4	51	204
2	52	104
4	54	216
4	55	220
3	56	168
4	57	228
1	58	58
1	60	60
3	61	183
1	62	62
2	64	128
1	65	65
1	68	68
1	69	69
42		2303

$$\frac{2303}{42} = 54.8$$

Therefore, the average president was about 54.8 years old at the time of inauguration.

The most frequently occurring number is called the *mode*. The mode can be determined from the frequency diagram by looking for the highest bar, although sometimes there will be ties. In the case of the age data, four numbers (51, 54, 55, and 57) all tie for the mode, since there were four presidents inaugurated at each of those ages.

For another example, on page 11 is a list of the areas (in square miles) of the 50 states.

1	Alaska	589,757	26	Wisconsin	56,154
2	Texas	267,338	27	Arkansas	53,104
3	California	158,693	28	North Carolina	52,586
4	Montana	147,138	29	Alabama	51,609
5	New Mexico	121,666	30	New York	49,576
6	Arizona	113,909	31	Louisiana	48,523
7	Nevada	110,540	32	Mississippi	47,716
8	Colorado	104,247	33	Pennsylvania	45,333
9	Wyoming	97,914	34	Tennessee	42,244
10	Oregon	96,981	35	Ohio	41,222
11	Utah	84,916	36	Virginia	40,817
12	Minnesota	84,068	37	Kentucky	40,395
13	Idaho	83,557	38	Indiana	36,291
14	Kansas	82,264	39	Maine	33,215
15	Nebraska	77,227	40	South Carolina	31,055
16	South Dakota	77,047	41	West Virginia	24,181
17	North Dakota	70,665	42	Maryland	10,577
18	Oklahoma	69,919	43	Vermont	9,609
19	Missouri	69,686	44	New Hampshire	9,304
20	Washington	68,192	45	Massachussetts	8,257
21	Georgia	58,876	46	New Jersey	7,836
22	Florida	58,560	47	Hawaii	6,450
23	Michigan	58,216	48	Connecticut	5,009
24	Illinois	56,400	49	Delaware	2,057
25	Iowa	56,290	50	Rhode Island	1,214

Putting the list in order does make it more meaningful than it otherwise would be. It would also help to draw a bar diagram. However, each number occurs only once in the list, so instead of drawing a bar diagram for each individual number we will group the states into different categories and count the number of states in each category:

Area range			States
0	to	10,000	8
10,000	to	20,000	1
20,000	to	30,000	1
30,000	to	40,000	3
40,000	to	50,000	8
50,000	to	60,000	9
60,000	to	70,000	3
70,000	to	80,000	3
80,000	to	90,000	4
90,000	to	100,000	2
100,000	to	110,000	1
110,000	to	120,000	2
120,000	to	130,000	1
140,000	to	150,000	1
150,000	to	160,000	1
260,000	to	270,000	1
580,000	to	590,000	1

Here is the frequency diagram corresponding to this chart (see Figure 2.3).

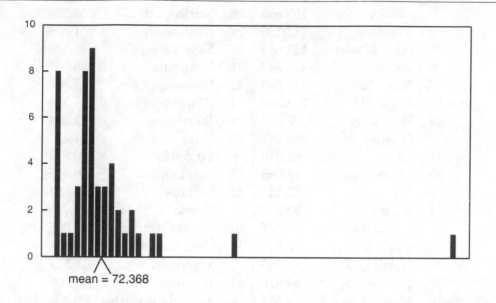

Figure 2.3

We can calculate the mean to be $\frac{3,618,400}{50} = 72,368$. The median is the average of the 25th and 26th states: $\frac{(56,290 + 56,154)}{2} = 56,222$. The mean is larger than the median, as we would expect, since there are a couple of very large states (Alaska, Texas) that pull up the mean.

Figure 2.4 illustrates three general types of frequency diagrams. If the frequencies are symmetric with a single mode, then the mean, median, and mode all coincide at the highest point, which is the middle of the distribution. The *normal distribution* is an example of a distribution like this; it will appear frequently in the rest of the book.

Some distributions are *bimodal*—that is, they have two modes. This could occur if we are asking people's opinions about a candidate who has a polarizing effect on the electorate. Quite a few people give the candidate a high rating; quite a few people give a low rating; and not many people are in the middle. If the distribution is uniformly symmetrically bimodal, as in the diagram, the mean and median coincide at the middle of the distribution, but neither of them is a very good indicator of a typical value.

Some distributions are asymmetrical with a long tail to the right. In this case the mean will be higher than the median, which will be higher than the mode.

VARIANCE AND STANDARD DEVIATION

We also want to measure the degree to which the numbers in a list are spread out. The measure commonly used for this purpose is called the *variance*. If the numbers in the list are all close to the mean, the variance will be small; if they are far away, the variance will be large.

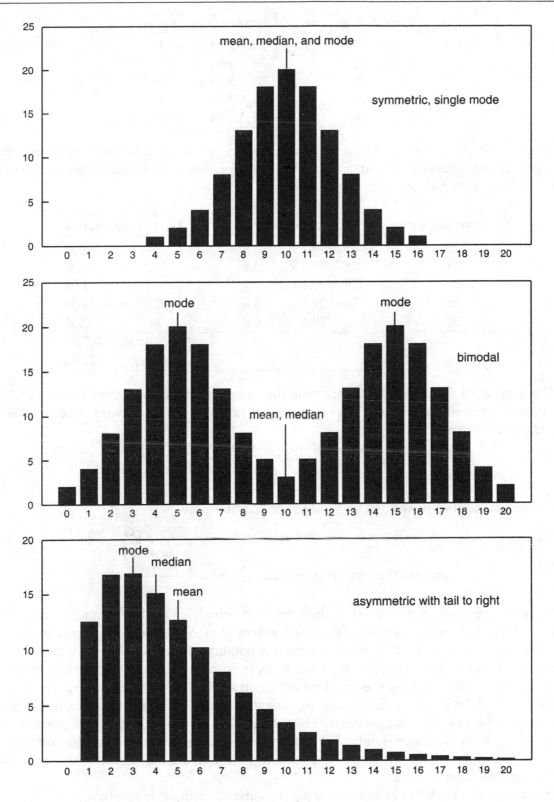

Figure 2.4

For the commuting data, we first find the distance from the mean for each element:

Commuting time	Distance from mean
25	$25 - 29.4 = -4.4$
38	$38 - 29.4 = 8.6$
28	$28 - 29.4 = -1.4$
26	$26 - 29.4 = -3.4$
30	$30 - 29.4 = 0.6$

We can't simply add these distances up and average them, because the negative numbers and positive numbers cancel out, giving zero. Therefore, we will square each of these numbers, making them all positive:

Commuting time	Distance from mean	Distance squared
25	$25 - 29.4 = -4.4$	$(-4.4)^2 = 19.36$
38	$38 - 29.4 = 8.6$	$8.6^2 = 73.96$
28	$28 - 29.4 = -1.4$	$(-1.4)^2 = 1.96$
26	$26 - 29.4 = -3.4$	$(-3.4)^2 = 11.56$
30	$30 - 29.4 = 0.6$	$0.6^2 = 0.36$
		Total: 107.2
		Average: 21.44

The average of the squared distance from the mean is defined to be the variance, 21.44 in this case. In general, the variance of a list of numbers $x_1, x_2, x_3, \ldots x_n$ is symbolized by $\mathrm{Var}(x)$, and is given by this formula:

$$\mathrm{Var}(x) = \frac{(x_1 - \overline{x})^2 + (x_2 - \overline{x})^2 + (x_3 - \overline{x})^2 + \ldots + (x_n - \overline{x})^2}{n}$$

This formula can also be written using summation notation:

$$\text{population variance} = \mathrm{Var}(x) = \sigma^2 = \frac{\sum_{i=1}^{n}(x_i - \overline{x})^2}{n}$$

The variance is also represented by the symbol σ^2, which is read as "sigma squared." The symbol σ is the lower case version of the Greek letter sigma. Make sure you don't confuse this with the upper case sigma (Σ) used for summation notation. The square root of the variance is called the *standard deviation*, and is represented by σ. Sometimes a subscript will be included if needed to specify what list the standard deviation applies to; for example, σ_x would refer to the standard deviation of the list x_1 to x_n. The standard deviation of the commuting times is $\sigma = \sqrt{21.44} = 4.630$. One advantage of the standard deviation is that it is measured in the same units as the original data. For example, the standard deviation of the commuting time is 4.63 minutes. The variance is measured in units that are the square of the units used to measure the original data.

In practice, it is slightly easier to calculate the variance using this formula:

$$\mathrm{Var}(x) = \overline{x^2} - \overline{x}^2$$

The quantity $\overline{x^2}$ is found by squaring each value of x, and then finding the average of those squares. This gives the following results:

Commuting time	x_i	x_i^2
25	$25^2 =$	625
38	$38^2 =$	1,444
28	$28^2 =$	784
26	$26^2 =$	676
30	$30^2 =$	900
	Total:	4,429
	Average:	885.8

Therefore, $\overline{x^2} = 885.8$. The variance is $885.8 - 29.4^2 = 885.8 - 864.36 = 21.44$, the same answer that was found using the longer formula.

The above formulas for variance and standard deviation apply when we have data taken from a population. When investigating data from a sample that is being used to estimate the properties of a population, then the variance is often calculated from a slightly different formula:

$$\text{sample variance} = s^2 = \frac{\sum_{i=1}^{n}(x_i - \overline{x})^2}{n-1}$$

The only difference is that $n-1$ appears in the denominator of the sample variance (which is often designated s^2). By contrast, n appears in the denominator of the population variance (designated by σ^2). The difference between the two versions of the variance will be discussed in Unit 6.

For more examples of descriptive statistics, see entries in Part II on **coefficient of variation; deciles; interquartile range; ogive; percentile; pie chart;** and **quartiles**.

EXERCISES

Each column below gives a list of numbers. For each list, calculate the mean (average), median, range (the difference between the highest and lowest values), variance, standard deviation, and the ratio of the standard deviation divided by the mean. When calculating the standard deviation, treat each list as a population, so use the version with n in the denominator.

1.	2.	3.	4.	5.	6.
4	65	97	18	62	83
10	60	86	12	61	83
6	52	75	16	62	87
3	52	89	20	61	82
7	52	43	17	62	84
10	60	83	27	60	850
6	66	76	30	60	82
2	66	33	21	61	82
9	51	67	24	60	80

7. Calculate the mean, median, and standard deviation for each of these quantities for the 50 states:

State	1980 Pop.	1990 Pop.	Area (mi^2)	Highest elevation (ft)	Electoral votes
Alabama	3,894	4,051	51,609	2,407	9
Alaska	402	550	589,757	20,320	3
Arizona	2,717	3,665	113,909	12,633	8
Arkansas	2,286	2,351	53,104	2,753	6
California	23,668	29,760	158,693	14,494	54
Colorado	2,890	3,294	104,247	14,433	8
Connecticut	3,108	3,287	5,009	2,380	8
Delaware	594	666	2,057	442	3
Florida	9,747	12,938	58,560	345	25
Georgia	5,463	6,478	58,876	4,784	13
Hawaii	965	1,108	6,450	13,796	4
Idaho	944	1,007	83,557	12,662	4
Illinois	11,427	11,431	56,400	1,235	22
Indiana	5,490	5,544	36,291	1,257	12
Iowa	2,914	2,777	56,290	1,670	7
Kansas	2,364	2,478	82,264	4,039	6
Kentucky	3,660	3,685	40,395	4,145	8
Louisiana	4,206	4,220	48,523	535	9
Maine	1,125	1,228	33,215	5,268	4
Maryland	4,217	4,781	10,577	3,360	10
Massachussetts	5,737	6,016	8,257	3,491	12
Michigan	9,262	9,295	58,216	1,980	18
Minnesota	4,076	4,375	84,068	2,301	10
Mississippi	2,521	2,573	47,716	806	7
Missouri	4,917	5,117	69,686	1,772	11
Montana	787	799	147,138	12,799	3
Nebraska	1,570	1,578	77,227	5,426	5
Nevada	801	1,202	110,540	13,143	4
New Hampshire	921	1,109	9,304	6,288	4
New Jersey	7,365	7,730	7,836	1,803	15
New Mexico	1,303	1,515	121,666	13,161	5
New York	17,558	17,990	49,576	5,344	33
North Carolina	5,880	6,629	52,586	6,684	14
North Dakota	653	639	70,665	3,506	3
Ohio	10,798	10,847	41,222	1,550	21
Oklahoma	3,025	3,146	69,919	4,973	8
Oregon	2,633	2,842	96,981	11,239	7
Pennsylvania	11,865	11,882	45,333	3,213	23
Rhode Island	947	1,003	1,214	812	4
South Carolina	3,121	3,487	31,055	3,560	8
South Dakota	691	696	77,047	7,242	3
Tennessee	4,591	4,877	42,244	6,643	11

Texas	14,225	16,987	267,338	8,749	32
Utah	1,461	1,723	84,916	13,528	5
Vermont	511	563	9,609	4,393	3
Virginia	5,347	6,187	40,817	5,729	13
Washington	4,132	4,867	68,192	14,410	11
West Virginia	1,950	1,793	24,181	4,863	5
Wisconsin	4,706	4,891	56,154	1,951	11
Wyoming	470	454	97,914	13,804	3

8. Calculate the mean age for the male population, female population, and overall population, given this frequency distribution. Assume the average age of all people less than 1 year old is 0.5; the average age of all 1-year-olds is 1.5; and so on. (Assume the average age of 85 to 89 year-olds is 87.5; the average age of those over 100 is 103.)

U.S. Population, 1992 (in thousands)

Age	Male	Female	Age	Male	Female	Age	Male	Female
<1	2,039	1,945	30	2,242	2,245	60	949	1,077
1	2,036	1,943	31	2,196	2,208	61	977	1,098
2	2,027	1,932	32	2,220	2,232	62	981	1,103
3	1,951	1,861	33	2,173	2,205	63	976	1,104
4	1,933	1,845	34	2,275	2,275	64	1,007	1,168
5	1,906	1,816	35	2,244	2,257	65	951	1,135
6	1,897	1,808	36	2,102	2,135	66	913	1,108
7	1,891	1,805	37	2,110	2,138	67	914	1,120
8	1,766	1,684	38	1,929	1,969	68	859	1,072
9	1,936	1,840	39	2,095	2,119	69	838	1,069
10	1,946	1,849	40	1,986	2,025	70	837	1,068
11	1,877	1,788	41	1,851	1,902	71	802	1,034
12	1,878	1,792	42	1,834	1,880	72	735	962
13	1,804	1,722	43	1,768	1,832	73	660	901
14	1,765	1,678	44	1,849	1,879	74	616	867
15	1,769	1,679	45	1,933	1,978	75	584	834
16	1,707	1,610	46	1,352	1,406	76	548	796
17	1,775	1,665	47	1,394	1,457	77	511	772
18	1,693	1,611	48	1,323	1,388	78	474	748
19	1,818	1,747	49	1,538	1,592	79	436	712
20	1,917	1,841	50	1,299	1,363	80	369	627
21	2,033	1,951	51	1,186	1,250	81	324	583
22	1,976	1,894	52	1,131	1,197	82	287	534
23	1,890	1,823	53	1,122	1,194	83	256	496
24	1,890	1,836	54	1,119	1,195	84	223	453
25	1,897	1,860	55	1,027	1,110	85–89	647	1,515
26	1,925	1,909	56	1,035	1,125	90–94	207	627
27	2,052	2,034	57	1,057	1,148	95–99	46	172
28	2,012	2,003	58	922	1,006	100+	10	35
29	2,255	2,243	59	981	1,076			

9. Calculate the mean income given the following 1990 income tax data. Assume the average income of the $1,000,000 and over category is $2,536,000, and treat the average of the less than $1,000 category as zero. For all other categories, use the midpoint as the average of the category.

Size of adjusted gross income		Number of returns (thousands)
Less than	1,000	3,688
1,000 –	3,000	7,379
3,000 –	5,000	6,317
5,000 –	7,000	6,004
7,000 –	9,000	6,026
9,000 –	11,000	5,891
11,000 –	13,000	5,573
13,000 –	15,000	5,382
15,000 –	17,000	4,686
17,000 –	19,000	4,656
19,000 –	22,000	6,308
22,000 –	25,000	5,465
25,000 –	30,000	7,838
30,000 –	40,000	12,283
40,000 –	50,000	8,837
50,000 –	75,000	10,944
75,000 –	100,000	3,276
100,000 –	200,000	2,330
200,000 –	500,000	644
500,000 –	1,000,000	130
	1,000,000 or more	61

ANSWERS

Here are the lists arranged in order:

	1.	2.	3.	4.	5.	6.
	10	66	97	30	62	850
	10	66	89	27	62	87
	9	65	86	24	62	84
	7	60	83	21	61	83
	6	60	76	20	61	83
	6	52	75	18	61	82
	4	52	67	17	60	82
	3	52	43	16	60	82
	2	51	33	12	60	80
Total	57	524	649	185	549	1,513
Mean, μ	6.3	58.2	72.1	20.6	61.0	168.1
Median	6	60	76	20	61	83
Range	8	15	64	18	2	770
Variance	7.78	37.95	404.77	28.47	0.67	58,124.77
σ	2.79	6.16	20.12	5.34	0.82	241.09
σ/μ	0.440	0.106	0.279	0.260	0.013	1.434

7.

	1980 Pop.	1980 Pop.	Area (mi^2)	Highest elevation (ft)	Electoral votes
Mean	4,518	4,962	72,368	6,162	10.7
Median	3,067	3,391	56,222	4,589	8.0
σ	4,668	5,405	87,391	5,035	9.4

8. male population: 34.00; female population: 36.82; overall population: 35.44

9. 31.51 thousand

UNIT 3

Probability

PROBABILITY OF AN EVENT

Draw one card from a well-shuffled standard 52-card deck. Are you more likely to draw the ace of hearts or the two of spades? You should be able to answer that question intuitively: those two events are equally likely. We will assume (without proof) that all 52 cards are equally likely to appear, and then develop some important rules of probability based on this fact.

What is the probability that you will draw a face card (a jack, queen, or king)? Figure 3.1 shows all 52 cards, with the 12 face cards shaded. We have 52 total outcomes; of these, 12 correspond to the event we're interested in (drawing a face card). Therefore, the probability of drawing a face card is 12/52.

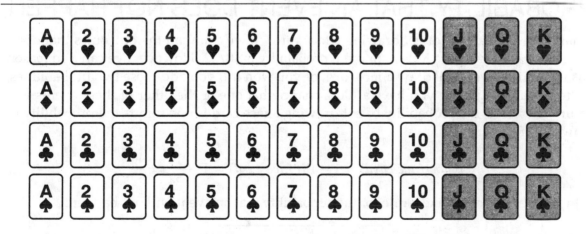

Figure 3.1: Pr(face card) = 12/52 = 3/13

In general, consider a situation where s represents the total number of outcomes, and all of those outcomes are equally likely. Let A represent some kind of interesting event consisting of a group of these outcomes. (In the example above, A represented the event of drawing a face card.) We need to count the number of outcomes corresponding to event A; call this number $N(A)$ (read "N of A"). Then we can calculate the probability that event A occurs with this formula:

$$\Pr(A) = \frac{N(A)}{s}$$

The expression "$\Pr(A)$" is read "the probability of event A."

What is the probability of drawing an ace? There are only four aces among the 52 cards, so the probability is 4/52.

What is the probability of drawing a red card (a diamond or a heart; see Figure 3.2)? There are 26 outcomes with red cards; therefore, the probability of a red card is $26/52 = 1/2$.

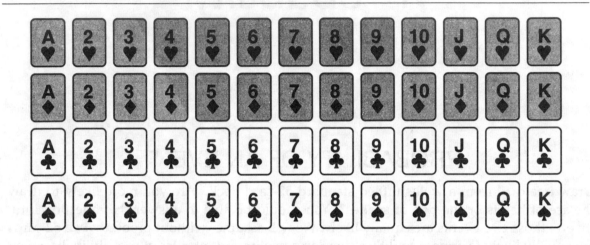

Figure 3.2: Pr(red card) = 26/52 = 1/2

PROBABILITY THAT AN EVENT DOES NOT HAPPEN

What is the probability of *not* drawing a face card? If we count all the outcomes that don't have face cards in Figure 3.1, we find the probability is 40/52. However, we can also reason this way: Either you will draw a face card, or you won't. If you add these two probabilities, the total will be 1, since you have a 100% chance that one or the other of these events will occur. Therefore, the probability of not drawing a face card is one minus the probability of drawing a face card:

$$\Pr(not\ face\ card) = 1 - \Pr(face\ card) = 1 - \frac{12}{52} = \frac{40}{52}$$

In general, we can make a rule for the probability that an event will not happen:

$$\boxed{\Pr(\text{not } A) = 1 - \Pr(A)}$$

The event "not A" is also symbolized $\tilde{A}$, or A_c, short for "A-complement."

PROBABILITY OF A OR B

What is the probability of drawing a face card or an ace? From Figure 3.3, we see there are 16 outcomes with either a face card or ace. Therefore:

$$\Pr(\text{face card or ace}) = \frac{16}{52}$$

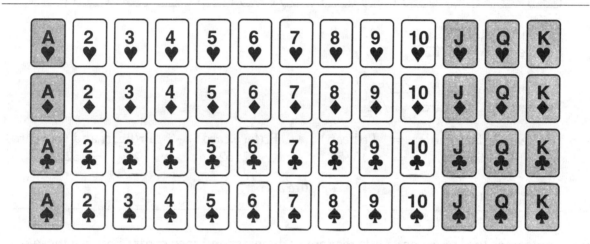

Figure 3.3: $\Pr(\text{face card or ace}) = \Pr(\text{face card}) + \Pr(\text{ace}) = 12/52 + 4/52 = 16/52 = 4/13$

Recall that $\Pr(\text{face card}) = 12/52$ and $\Pr(\text{ace}) = 4/52$. We can add these together:

$$\frac{12}{52} + \frac{4}{52} = \frac{16}{52}$$

Perhaps we have stumbled across a general rule. We are tempted to ask: Does $\Pr(A \text{ or } B) = \Pr(A) + \Pr(B)$?

However, a glance at Figure 3.4 quickly throws cold water on that idea. We can see that the probability of drawing a heart is 13/52, and we know the probability of drawing a face card is 12/52. Suppose we want to find probability of drawing either a face card or a heart. We can count 22 outcomes with a face card or a heart, so the probability of this happening is 22/52. However, if we add together the probabilities of these two individual events, we have:

$$\frac{13}{52} + \frac{12}{52} = \frac{25}{52}$$

This does not equal 22/52. What went wrong with our rule that said we could add the probabilities?

Further analysis of Figure 3.4 reveals the problem. If we add together the number of outcomes with hearts and the number of outcomes with face cards, we are counting twice those

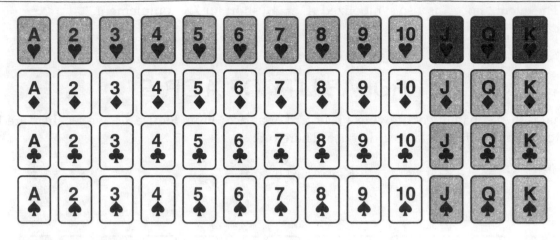

Pr(heart) = 13/52
Pr(face card) = 12/52

Pr(heart or face card) = 22/52
Pr(heart and face card) = 3/52

Pr(heart or face card)	=	Pr(heart)	+	Pr(face card)	−	Pr(heart and face card)
22/52	=	13/52	+	12/52	−	3/52

Figure 3.4

three outcomes that represent cards that are both hearts and face cards. Because these outcomes have been counted twice, we have to subtract them once in our formula for Pr(A or B):

$$Pr(A \text{ or } B) = Pr(A) + Pr(B) - Pr(A \text{ and } B)$$

There is no possibility of drawing a card that is both an ace and a face card (review Figure 3.3). These two events are said to be *mutually exclusive*. If one of them occurs, we know the other one cannot occur. Another way of saying it is this:

A and B are mutually exclusive events if Pr(A and B) = 0.

With mutually exclusive events, we have no need to worry about the problem of counting twice, so we can use the simple addition rule for Pr(A or B). If the events are not mutually exclusive, then we have to use the general rule:

If A and B are mutually exclusive:	Pr(A or B) = Pr(A) + Pr(B)
In general:	Pr(A or B) = Pr(A) + Pr(B) − Pr(A and B)

Now we embark upon the quest for a general formula for Pr(A and B). If we can list all the probabilities, as in Figure 3.4, we can simply count:

$$Pr \text{ (heart and face card)} = \frac{3}{52}$$

It would be helpful to have a general rule that would work in other cases. We have already seen what would happen if you added together the probabilities of two events. You might wonder: Perhaps we should try multiplying the probabilities?

$$\text{Pr (heart)} \times \text{Pr (face card)} = \frac{13}{52} \times \frac{12}{52}$$

Multiplying these two together gives the fraction 156/2704, but it would be easier to see what happens if we write the numerator and denominator as the product of the prime factors:

$$\frac{13}{52} \times \frac{12}{52} = \frac{13 \times 3 \times 2^2}{2^2 \times 13 \times 2^2 \times 13}$$

We now have the advantage that we can start canceling:

$$\frac{13 \times 3 \times 2^2}{2^2 \times 13 \times 2^2 \times 13} = \frac{3}{13 \times 2^2} = \frac{3}{52}$$

This is the same as Pr(heart and face card) from Figure 3.4. This gives us hope that we might have discovered a rule: Does $\text{Pr}(A \text{ and } B) = \text{Pr}(A) \times \text{Pr}(B)$?

Once again our hopes are quickly dashed. This cannot be a general rule, since we recall that $\text{Pr}(A \text{ and } B) = 0$ if A and B are mutually exclusive. Here is another example where the proposed rule breaks down:

$$\text{Pr(king and face card)} = \frac{4}{52} \neq$$

$$\text{Pr(king)} \times \text{Pr(face card)} = \frac{4}{52} \times \frac{12}{52} = \frac{48}{2704}$$

See Figure 3.5.

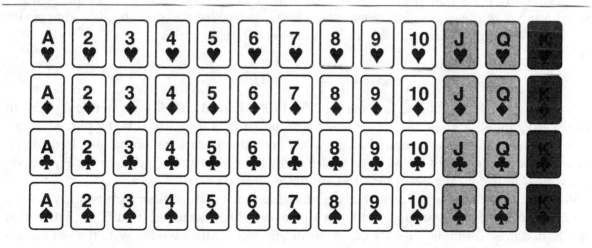

Figure 3.5: Pr(king | face card) = 4/12 = 1/3

There must be some cases where $\text{Pr}(A \text{ and } B) = \text{Pr}(A) \times \text{Pr}(B)$; we need to determine what those cases are.

CONDITIONAL PROBABILITY

Imagine playing a card game where it's legal to cheat—a little bit. You need to guess whether or not the card on the top of the deck is a king. Ordinarily, you know:

$$\Pr(\text{king}) = \frac{4}{52}$$

However, in this game you are allowed a little peek at the card. This peek is not enough to tell you what the card is, but you are able to tell that the card is a face card. Suddenly the situation changes; now we know the probability of it being a king is 4/12—there are 12 face cards, and since we know the card must be a face card these are the only possible outcomes.

"How can the probability change like that?" you might object. "The card itself hasn't changed just because you took your little peek at it. It either is a king or not, whether or not you looked at it."

This objection illustrates an important feature of probability: the probability of an event is not just an inherent property of the event; probability is always in some way a measure of our own knowledge and our own ignorance. In any situation, we have some information and we lack other information. Probabilities are always based upon what we know, so if our information changes, the probability changes. Suppose you actually looked at the card and saw that it was a king; then the probability of it being a king changes to 1. There is nothing about the nature of the card that changed, but changing our information changes the probability.

We will write it this way: The *conditional probability* that we will draw a king, *given that* we have drawn a face card, is 4/12. In symbols, this is written:

$$\Pr(\text{king} \mid \text{face card}) = \frac{4}{12} = \frac{1}{3}$$

The vertical line "|" is read "given that." A conditional probability applies to a situation where knowing that one event occurs (in this case, the event of drawing a face card) changes the probability that another event occurs (drawing a king). The event of drawing a king is called a *subset* of the event of drawing a face card, since every outcome that is a king is also a face card. In general, if event A is a subset of event B, then:

$$\Pr(A \mid B) = \frac{\Pr(A)}{\Pr(B)}$$

Now we look for a general formula for $\Pr(A \mid B)$, where A and B are *any* two events. In some card games we are able to observe some of the cards as the game is being played, so we know these cards are no longer in the deck. Therefore, the conditional probability of drawing a particular card changes based upon which cards have already been drawn. In Figure 3.6, let event B represent the event that the card drawn is in the shaded area.

We want to determine the probability of drawing a jack. The ordinary probability is 4/52 = 1/13. However, now suppose we know that event B has occurred (that is, we know that the shaded area contains the only cards left in the deck). Since we know that event B has occurred, there are only 14 possible outcomes; of these, only 3 represent jacks. Therefore, we find $\Pr(\text{jack} \mid B) = 3/14$.

In general, we find the conditional probability that event A occurs, given that event B occurs, from this formula:

$$\Pr(A \mid B) = \frac{\Pr(A \text{ and } B)}{\Pr(B)}$$

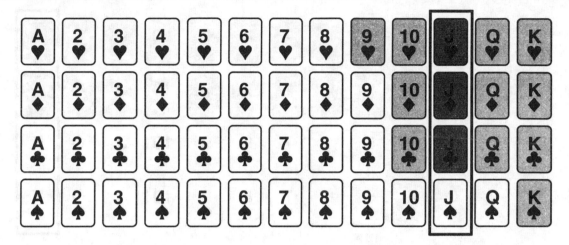

$B =$ event card is in shaded area
$\Pr(B) = 14/52$
$\Pr(\text{jack}) = 4/52$
$\Pr(\text{jack and } B) = 3/52$
$\Pr(\text{jack} \mid B) = 3/14$

Figure 3.6

Here are two important examples that illustrate general rules:

- What is the probability of drawing a king, given that the card is a number card? The answer here is obviously zero. These two events (drawing a king and drawing a number card) are mutually exclusive. In general, if A and B are mutually exclusive events, then $\Pr(A \text{ and } B)$ is zero, so $\Pr(A \mid B)$ is also zero.

- What is the probability of drawing a king, given that the card is a red card (See Figure 3.7)? We can see that $\Pr(\text{red card}) = 26/52 = 1/2$, and $\Pr(\text{king and red card}) = 2/52$. Therefore:

$$\Pr(\text{king} \mid \text{red card}) = \frac{\frac{2}{52}}{\frac{26}{52}} = \frac{2}{26} = \frac{1}{13}$$

The interesting fact we notice here is that the conditional probability of drawing a king given a red card is the same as the original probability of drawing a king—without knowing the color of the card. Suppose a foolish cheater rigged a device allowing him to see the color of the card, but not the type of card. If he then used this device to try to predict whether or not he would draw a king, he would have utterly no advantage over the honest player without that information. In a case such as this, where knowledge of one event does not help you know whether another event occurs, the two events are said to be *independent*. The general definition is: A and B are independent events if and only if $\Pr(A \mid B) = \Pr(A)$.

We can derive an important consequence from this definition. Recall the definition of conditional probability:

$$\Pr(A \mid B) = \frac{\Pr(A \text{ and } B)}{\Pr(B)}$$

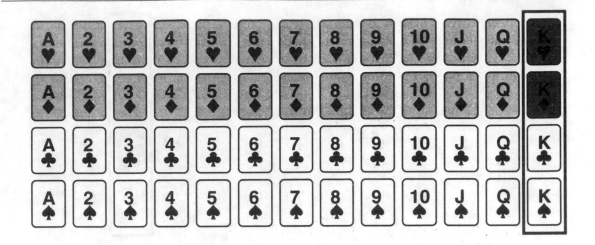

Figure 3.7: Pr(king | red card) = 2/26 = 1/13

In the case of independent events, we can substitute Pr(A) in place of Pr($A \mid B$):

$$\Pr(A) = \frac{\Pr(A \text{ and } B)}{\Pr(B)}$$

We can rewrite this:

$$\Pr(A \text{ and } B) = \Pr(A) \times \Pr(B)$$

Now we have discovered when it is all right to multiply two probabilities: when the two events are independent, multiplying the probabilities gives you the probability that both events occur.

Conditional Probability Summary
General definition:
$$\Pr(A \mid B) = \frac{\Pr(A \text{ and } B)}{\Pr(B)}$$
Pr($A \mid B$) is read "the probability that event A will occur, given that event B has occurred," or, more concisely, "the probability of A given B."

If A and B are mutually exclusive events:
Pr($A \mid B$) = 0

If A is a subset of B:
$$\Pr(A \mid B) = \frac{\Pr(A)}{\Pr(B)} \quad \Pr(A \text{ and } B) = \Pr(A)$$

If A and B are independent events:
Pr($A \mid B$) = Pr(A)
Pr(A and B) = Pr(A) × Pr(B)

EXAMPLE 1

You have just rolled a 5 on a die. What is the probability that the next die you roll will show 5?

Since the two events are independent, the probability of rolling a 5 on the second die is 1/6—the same as it would be if you didn't know what the first die was.

EXAMPLE 2

You have vitally important data on your computer disk. You estimate there is a .01 probability that the system will fail on any day, causing a catastrophic loss of data. You realize you need a backup data system, which will also have a .01 probability of failing. What is the probability that both systems will fail?

If the two systems are independent, then you can simply multiply the probabilities:

$$\text{Pr(main system fails and backup fails)} =$$
$$\text{Pr(main system fails)} \times \text{Pr(backup fails)} = .01 \times .01 = .0001$$

You can sleep easier with this knowledge, but if this probability is still too high you could add a third system:

$$\text{Pr (all three systems fail)} = .01^3 = .000001$$

which is one in a million.

It is vital to remember, however, the importance of the assumption of independence. Suppose your backup system is located next to the main system. In that case, a single bad event could disable both systems. Therefore, it would not be correct to treat the systems as independent, and the probability of a double failure is much greater than indicated above. The best strategy would be to have the backup system in a different building, so it would be more independent of the main system. Perhaps you should have it in a different city if you are worried about a city-wide disaster. However, you have to balance the costs of maintaining the backup with the expected benefits. It doesn't help to be too paranoid. You could argue that the backup should be on a different planet, since a giant asteroid could strike the earth. This illustrates that you can never have these systems be absolutely independent, but you should have them be as independent as is economical.

This example illustrates another feature of probability. We did not calculate the probability of failure (.01) for one system by listing all outcomes and then counting the ones with failure. Instead, this kind of probability is only an estimate. It will never be as clear-cut as probabilities involving games of chance are. Nevertheless, we assume that we can use the same rules of probability that apply in games of chance to other situations, even where the probabilities are just estimated values.

COUNTING OUTCOMES: THE MULTIPLICATION PRINCIPLE

Suppose your computer system is protected by a seven-letter password. What is the probability that an intruder could come up with the correct password in one random guess?

Recall the formula for probability: $\Pr(A) = \frac{N(A)}{s}$. In this case $N(A) = 1$, since there is only one outcome with the correct password. To find the probability, all we need to do is find s, the total number of possible passwords. In our previous example, we simply listed all the outcomes. We could start doing that:

$$\text{AAAAAAA, \quad AAAAAAB, \quad AAAAAAC, \quad AAAAAAD, ...}$$

But we quickly realize a problem: there are far too many outcomes to attempt to list them all. Fortunately, we don't need the complete list to solve the problem—we just need to know how many possibilities there are. In order to solve probability problems, we need to develop an important skill, the ability to calculate the number of outcomes in a given situation without having to list them all. There are a few general formulas that apply to some common situations.

While thinking about the password problem, let's visit a restaurant that offers 4 choices for main dishes (chicken, beef, ham, and fish) and 3 choices for side dishes (soup, salad, and rice). If we eat at the restaurant every day, how many days will we be able to order different meals before we have to start repeating our choices?

It is not too hard to list all of the possibilities:

chicken/soup	chicken/salad	chicken/rice
beef/soup	beef/salad	beef/rice
ham/soup	ham/salad	ham/rice
fish/soup	fish/salad	fish/rice

Each of the 4 main dishes can be matched with any one of the 3 side dishes, so there are $4 \times 3 = 12$ possibilities in all.

We can state this idea as a general rule:

> **Multiplication Principle I**
> If k items from one list can be combined with m items from another list, there are km possibilities.

SAMPLING WITH REPLACEMENT

Suppose there were only two letters in the password. We reason there are 26 possibilities for the first letter; each of these possibilities can be matched with any one of the 26 possibilities for the second letter. Therefore, there are $26^2 = 676$ possible two-letter passwords. Your chance of guessing correctly is $\frac{1}{676} = .00148$. Extending this reasoning, there are $26^3 = 17,576$ possible three-letter passwords; $26^4 = 456,976$ possible four-letter passwords, and

$26^7 = 8,031,810,176$ possible seven-letter passwords. Therefore, the chance of randomly guessing a seven-letter password is $1/8{,}031{,}810{,}176 = 1.245 \times 10^{-10}$. (Remember, this only applies if the password has been selected randomly, and if the intruder has only one chance to guess.) We have been assuming that a letter can occur more than once in the password. The term *sampling with replacement* is used to refer to a situation where an item can be selected more than once. You can imagine that you choose the letters from a table full of wooden blocks containing all 26 letters; after you have chosen one and written it down, you replace the block on the table so that letter can be chosen again. (Later in this unit we will consider situations of sampling without replacement.)

Multiplication Principle II
If you make n choices with replacement from a group of m objects, the total number of possibilities is m^n.

EXAMPLE 3

If you toss ten coins, what is the probability they will all be heads?

With one coin, there are 2 possibilities; with two coins, there are $2^2 = 4$ possibilities (HH, HT, TH, TT); with three coins there are $2^3 = 8$ possibilities (HHH, HHT, HTH, HTT, THH, THT, TTH, TTT). These eight possibilities can be illustrated with a diagram called a *tree diagram* (see Figure 3.8).

With ten coins, there are $2^{10} = 1{,}024$ possibilities. Only one of these consists of all heads, so the probability is $1/1{,}024 = .000977$.

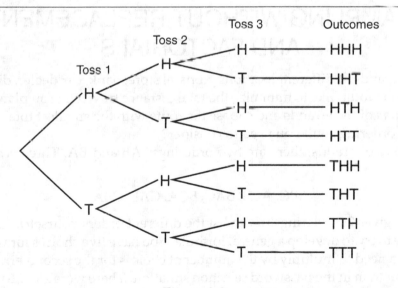

Figure 3.8

EXAMPLE 4

If you toss five dice, what is the probability that all five dice will show six?

With one die, there are 6 possibilities. With two dice, there are $6^2 = 36$ possibilities (see Figure 3.9). In general, with n dice there are 6^n possibilities. If you toss five dice, there are $6^5 = 7,776$. Only one of these possibilities shows all sixes, so the probability of this happening is $1/7,776 = .000129$. Alternatively, if we were interested in the probability that all five dice will be the same, the result is $6/7,776 = .000772$ (since there are six possibilities where all five dice are the same).

6 choices for number on first die	6 choices for number on second die					
	1	2	3	4	5	6
1	(1, 1)	(1, 2)	(1, 3)	(1, 4)	(1, 5)	(1, 6)
2	(2, 1)	(2, 2)	(2, 3)	(2, 4)	(2, 5)	(2, 6)
3	(3, 1)	(3, 2)	(3, 3)	(3, 4)	(3, 5)	(3, 6)
4	(4, 1)	(4, 2)	(4, 3)	(4, 4)	(4, 5)	(4, 6)
5	(5, 1)	(5, 2)	(5, 3)	(5, 4)	(5, 5)	(5, 6)
6	(6, 1)	(6, 2)	(6, 3)	(6, 4)	(6, 5)	(6, 6)

Figure 3.9

SAMPLING WITHOUT REPLACEMENT
AND FACTORIALS

You have five errands at different locations: apparel store, bank, car dealer, dime store, electronics store. You would like to minimize the total distance traveled. You plan to consider all possible orderings of the errands and choose the route with the smallest total distance. How many different orderings will you need to consider?

If there were two errands, there are two orderings: AB and BA. Three errands gives six orderings:

ABC, ACB, BAC, BCA, CAB, CBA

Four errands gives 24 orderings (try to list the different orders yourself).

Now it's time to try to develop a general formula. You have five choices for which errand to do first. We then need to multiply by the number of choices for the second errand, but things are now different than in the password selection situation. There were 26×26 possibilities for the first two letters of the password, because the second letter could be the same as the first one (that was called sampling with replacement). However, we will not do the same errand twice, so there are only 4 choices left for the second errand. Therefore, there are $5 \times 4 = 20$ choices for the first two errands. This type of choice is called *sampling without replacement*, since once an item has been selected it cannot be selected again. Only 3 choices remain for the third errand, so there are $5 \times 4 \times 3 = 60$ ways of choosing the first three errands. There are

$5 \times 4 \times 3 \times 2 = 120$ possible ways of choosing the first four errands. Once you have decided on the first four, however, there is only one choice left for the last errand, so we will write the answer as $5 \times 4 \times 3 \times 2 \times 1 = 120$.

This number is small enough to list all of the possibilities, which is good because it helps to be able to see a complete list to visualize the patterns. Unfortunately, we will lose this help when the number of possibilities becomes too large to list. Note that the table below is arranged in five columns, each corresponding to a different choice for the first errand; each column contains four blocks, representing different choices for the second errand; each block contains three pairs, representing different choices for the third errand.

AB CDE	BA CDE	CA BDE	DA BCE	EA BCD
AB CED	BA CED	CA BED	DA BEC	EA BDC
AB DCE	BA DCE	CA DBE	DA CBE	EA CBD
AB DEC	BA DEC	CA DEB	DA CEB	EA CDB
AB ECD	BA ECD	CA EBD	DA EBC	EA DBC
AB EDC	BA EDC	CA EDB	DA ECB	EA DCB
AC BDE	BC ADE	CB ADE	DB ACE	EB ACD
AC BED	BC AED	CB AED	DB AEC	EB ADC
AC DBE	BC DAE	CB DAE	DB CAE	EB CAD
AC DEB	BC DEA	CB DEA	DB CEA	EB CDA
AC EBD	BC EAD	CB EAD	DB EAC	EB DAC
AC EDB	BC EDA	CB EDA	DB ECA	EB DCA
AD BCE	BD ACE	CD ABE	DC ABE	EC ABD
AD BEC	BD AEC	CD AEB	DC AEB	EC ADB
AD CBE	BD CAE	CD BAE	DC BAE	EC BAD
AD CEB	BD CEA	CD BEA	DC BEA	EC BDA
AD EBC	BD EAC	CD EAB	DC EAB	EC DAB
AD ECB	BD ECA	CD EBA	DC EBA	EC DBA
AE BCD	BE ACD	CE ABD	DE ABC	ED ABC
AE BDC	BE ADC	CE ADB	DE ACB	ED ACB
AE CBD	BE CAD	CE BAD	DE BAC	ED BAC
AE CDB	BE CDA	CE BDA	DE BCA	ED BCA
AE DBC	BE DAC	CE DAB	DE CAB	ED CAB
AE DCB	BE DCA	CE DBA	DE CBA	ED CBA

It turns out that in probability we frequently need to calculate the product of all whole numbers from one up to a given number. It helps to give a name to this quantity. We call it the *factorial* of that number, and represent it by an exclamation point (!). Think of the exclamation point as representing the fact that surprisingly large numbers can be generated this way.

> **Factorial**
> The factorial of a number n (written $n!$) is the product of
> all whole numbers from one up to that number.
> $0! = 1$ (by definition)
> $1! = 1$
> $2! = 2 \times 1 = 2$
> $3! = 3 \times 2 \times 1 = 6$
> $4! = 4 \times 3 \times 2 \times 1 = 24$
> $5! = 5 \times 4 \times 3 \times 2 \times 1 = 120$
> $6! = 6 \times 5 \times 4 \times 3 \times 2 \times 1 = 720$
> $7! = 7 \times 6 \times 5 \times 4 \times 3 \times 2 \times 1 = 5,040$
> $8! = 8 \times 7 \times 6 \times 5 \times 4 \times 3 \times 2 \times 1 = 40,320$
> $9! = 9 \times 8 \times 7 \times 6 \times 5 \times 4 \times 3 \times 2 \times 1 = 362,880$
> $10! = 10 \times 9 \times 8 \times 7 \times 6 \times 5 \times 4 \times 3 \times 2 \times 1 = 3,628,800$
> If you have n distinct objects, then $n!$ gives the number
> of ways of putting the objects in different orders.

You might object that the definition $0! = 1$ seems strange and unnatural, but our reasoning is that if you have zero objects, there is only one way to put them in order. You could raise a philosophical objection to the idea of putting 0 objects in any kind of order, but since the definition $0! = 1$ makes some future formulas work out well we will stick with it.

EXAMPLE 5

An indecisive baseball manager decides to try all possible batting orders before deciding which is best. Since there are 9 players, there are $9! = 362,880$ different orders. It will take a very long time to try them all.

EXAMPLE 6

Suppose you will try to guess the exact order of all 52 cards in a well-shuffled deck. There are $52! = 8.066 \times 10^{67}$ different orderings for the 52 cards, so the probability of guessing the correct order is $1/8.066 \times 10^{67} = 1.240 \times 10^{-68}$.

PERMUTATIONS

What is the probability you can correctly guess the order of the top three finishers in an eight-horse race? There are $8! = 40,320$ possible finish orders for the eight horses, so your chances of guessing the order of all horses is $1/40,320 = 2.48 \times 10^{-5}$. However, the situation is not that grim, because we don't need to guess the order for all horses—just the top three. There are 8 possibilities for the first place horse. For each of these, there are 7 horses left that might finish second, giving $8 \times 7 = 56$ ways of selecting the top two horses. For each of these there are 6 horses left that might finish third, so there are $8 \times 7 \times 6 = 336$ possiblities for the top three horses. We will not attempt to list all of them, so now it is time to take the leap of faith and trust that we truly can find formulas to calculate the number of outcomes without having to list them all. There is a $1/336 = .00298$ chance that you could guess the order of the top three finishers.

The term *permutations* is used for the 336 possible finishes in the horse race. We can write the expression this way:

$$8 \times 7 \times 6 = \frac{8 \times 7 \times 6 \times 5 \times 4 \times 3 \times 2 \times 1}{5 \times 4 \times 3 \times 2 \times 1}$$

"That did not help!" you vehemently object, since the new formula is more complicated than the original one. However, we now have both the top and the bottom of the fraction written in a form where we can use the factorial notation:

$$8 \times 7 \times 6 = \frac{8!}{5!}$$

Here is the general definition:

Permutations

If you select j objects without replacement from a group of n objects,
and you do wish to count each ordering separately, then the number of ways of making the selections is given by:

$$\frac{n!}{(n-j)!}$$

This formula gives the number of permutations of n objects taken j at a time.

In the horse race example, $n = 8$ and $j = 3$, giving us $8!/(8-3)! = 8!/5!$.

COMBINATIONS

As an ice cream connoisseur, you only patronize ice cream parlors that offer a rich variety of ice cream experiences. You are testing out one establishment that offers two-scoop cones chosen from among 30 flavors. The owner, knowing about the formula for permutations, promises $30!/(30-2)! = 30!/28! = 30 \times 29 = 870$ distinctly different two-scoop cones. You decide to try them all, but you soon are in for a bitter surprise when you are served a cone with vanilla on top, chocolate on the bottom.

"You already served me a vanilla/chocolate cone!" you protest to the owner.

"I beg to differ!" the owner responds. "I have previously served you a chocolate on top, vanilla on bottom cone. You can plainly see that today I am serving you vanilla on top, chocolate on bottom."

"I don't care which one is on top!" you respond. "I only care about which flavors are chosen."

You understand there are 870 permutations of 30 objects selected 2 at a time, but with dismay you realize that in the permutations formula each ordering of the objects is counted separately. Therefore, chocolate/vanilla is counted separately from vanilla/chocolate. Likewise, every possible pair of flavors is counted twice in the permutations formula, representing the two different orderings. If you don't care about the order, then there are only 870/2 = 435 possibilities. The formula we use in this case is called the *combinations* formula. The number of combinations is equal to the number of permutations divided by the number of orderings of the chosen objects. In the ice cream example, the number of permutations is:

$$\frac{30!}{28!}$$

Divide this by 2! = 2 (the number of orderings) to get the number of combinations:

$$\frac{30!/28!}{2!} = \frac{30!}{2!\,28!}$$

In general, we have found that there are $j!$ different orderings of j objects, so the final formula for combinations works out to be:

Combinations
If you select j objects without replacement from a group of n objects, and you do *not* wish to count each ordering separately, then the number of ways of making the selections is given by:
$$\frac{n!}{j!(n-j)!}$$
This formula gives the number of combinations of n objects taken j at a time. The formula is often represented with this notation:
$$\binom{n}{j} = \frac{n!}{j!(n-j)!}$$

Note: the formula can be written either

$$\binom{n}{j} = \frac{n!}{j!(n-j)!} \quad \text{or} \quad \binom{n}{j} = \frac{n!}{(n-j)!j!}$$

EXAMPLE 7

A pizza restaurant allows you to choose three different toppings for your pizza, selected from a list of five choices: anchovies, bacon, cheese, diced ham, and eggplant. How many different choices do you have?

From the permutations formula, with $n = 5$ and $j = 3$, we have $5!/2! = 5 \times 4 \times 3 = 60$ permutations, as shown on the following page:

ABC	ACB	BAC	BCA	CAB	CBA
ABD	ADB	BAD	BDA	DAB	DBA
ABE	AEB	BAE	BEA	EAB	EBA
ACD	ADC	CAD	CDA	DAC	DCA
ACE	AEC	CAE	CEA	EAC	ECA
ADE	AED	DAE	DEA	EAD	EDA
BCD	BDC	CBD	CDB	DBC	DCB
BCE	BEC	CBE	CEB	EBC	ECB
BDE	BED	DBE	DEB	EBD	EDB
CDE	CED	DCE	DEC	ECD	EDC

Note that every entry in each row consists of the same three letters, listed six different times in each of the six different orderings. You don't really care about the number of permutations, since the order in which the toppings are placed on the pizza doesn't matter. Therefore, we need to look at the number of combinations:

$$\binom{5}{3} = \frac{5!}{3!2!} = \frac{5 \times 4 \times 3 \times 2 \times 1}{3 \times 2 \times 1 \times 2 \times 1}$$

After canceling:

$$\frac{5 \times 4 \times 3}{3 \times 2 \times 1} = \frac{60}{6} = 10$$

We can list the ten different combinations:

ABC, ABD, ABE, ACD, ACE, ADE, BCD, BCE, BDE, CDE

Note that the number of combinations (10) is much smaller than the number of permutations (60).

EXAMPLE 8

Five cards will be dealt from a 52-card deck. What is the probability that you can guess which five cards will be drawn? We are selecting $j = 5$ objects from a group of $n = 52$ objects. Now we wonder: Do we use the permutations formula or the combinations formula? Since we don't care about the order in which the cards are dealt to us (we only care about which cards we get), we use the combinations formula:

$$\binom{52}{5} = \frac{52!}{5!47!}$$

The 47! cancels out most of the top of the fraction, sparing us from the labor of writing that all out, and leaving us with:

$$\binom{52}{5} = \frac{52 \times 51 \times 50 \times 49 \times 48 \times 47!}{5 \times 4 \times 3 \times 2 \times 1 \times 47!} = 2,598,960$$

Therefore, our chance of guessing the cards correctly is $1/2{,}598{,}960 = 3.848 \times 10^{-7}$.

The combinations formula plays an important role in sampling. See Unit 5, where these ideas are developed further. See also **combinations** and **hypergeometric distribution** in Part II.

EXERCISES

1. (a) If you toss a coin five times, what is the probability that all five tosses will be heads? Give a formula and a numerical result. (b) If you toss a coin 5 times, what is the probability that at least one of the tosses will be a head? Give a formula and a numerical result.

2. (a) If you toss two dice, what is the probability that at least one six will appear? (b) If you toss n dice, what is the probability that at least one six will appear?

3. Suppose there is a 68% chance that there will be traffic congestion on the North Bridge at 5 P.M., a 62% chance that there will be traffic congestion on the South Bridge, and a 52% chance that there will be traffic congestion on both bridges. What is the probability that there will be traffic congestion on at least one bridge?

---

4. Suppose your factory contains two machines. There is a 7% probability that the first machine will break down on a particular day, and a 10% probability that the second machine will break down on a particular day. The two machines are independent—that is, whatever happens to one of the machines has no effect on the other machine. Show the formulas you use for this problem. (a) What is the probability that both machines will break down? (b) What is the probability that at least one machine will break down? (c) What is the probability that one (but only one) of the machines will break down?

---

5. A toolbox contains four hammers, three wrenches, and two screwdrivers. In each case give both a formula and a numeric result. (a) If three tools are selected at random, what is the probability that all three tools will be hammers? (b) If three tools are selected at random, what is the probability that one of each type of tool will be selected? (c) If seven tools are selected at random, what is the probability that no screwdrivers will be selected? (d) If six tools are selected at random, what is the probability that two of each type of tool will be selected?

6. Suppose you roll three dice. List the possible ways of rolling a 10 as the sum of the numbers on the three dice, and calculate the probability of obtaining a 10. Hint: First, list the number of possible ways where you do not count different orderings of the dice separately (for example, 1, 3, 6 would be counted the same as 6, 3, 1). Then use that information to calculate the total number of different ways of rolling a 10.

ANSWERS

1. (a) $(1/2)^5 = .03125$
(b) The probability of obtaining at least one head is equal to one minus the probability of obtaining all tails:

$$1 - .031 = .969$$

2. (a) Let A = event of rolling a 6 on the first die
Let B = event of rolling a 6 on the second die
$Pr(A) = 1/6$; $Pr(B) = 1/6$
$Pr(A$ and $B) = 1/36$ (since they are independent)

$$
\begin{aligned}
Pr(A \text{ or } B) &= Pr(A) + Pr(B) - Pr(A \text{ and } B) \\
&= 1/6 + 1/6 - 1/36 \\
&= 11/36
\end{aligned}
$$

(b) Let A = event of not rolling a 6 on the first die; $Pr(A) = 5/6$
Let B = event of not rolling a 6 on the second die; $Pr(B) = 5/6$
$Pr(A$ and $B) = 25/36$ (since they are independent)
By the same reasoning, the probability of rolling no sixes on n tosses is $(5/6)^n$. The probability of rolling at least one six is equal to one minus the probability of rolling no sixes:

$$1 - (5/6)^n$$

3. Let A = event that North Bridge is congested
Let B = event that South Bridge is congested

$$
\begin{aligned}
Pr(A \text{ or } B) &= Pr(A) + Pr(B) - Pr(A \text{ and } B) \\
&= .68 + .62 - .52 \\
&= .78
\end{aligned}
$$

(Note that these two events are not independent; congestion on one bridge is more likely to occur if the other bridge is congested.)

4. Let A = event machine 1 breaks down
Let B = event machine 2 breaks down
(a) $Pr(A$ and $B) = .07 \times .10 = .007$
(since the two events are independent)

(b)
$$
\begin{aligned}
Pr(A \text{ or } B) &= Pr(A) + Pr(B) - Pr(A \text{ and } B) \\
&= .07 + .10 - .007 \\
&= .163
\end{aligned}
$$

(c) The probability that exactly one machine will break down is equal to the probability that at least one machine breaks down minus the probability that they both break down:
$.163 - .007 = .1560$

5. (a) There are $\binom{9}{3} = 84$ ways to select the three tools. There are $\binom{4}{3} = 4$ ways to select three tools that are all hammers. The probability of this is therefore $\frac{4}{84} = .048$.

(b) There are $\binom{4}{1}$ ways to select one hammer, $\binom{3}{1}$ ways to select one wrench, and $\binom{2}{1}$ ways to select one screwdriver. Therefore, the probability is

$$\frac{\binom{4}{1} \times \binom{3}{1} \times \binom{2}{1}}{\binom{9}{3}} = \frac{24}{84} = .286$$

(c)

$$\frac{1}{\binom{9}{7}} = \frac{1}{36} = .028$$

(d)

$$\frac{\binom{4}{2} \times \binom{3}{2} \times \binom{2}{2}}{\binom{9}{6}} = \frac{6 \times 3 \times 1}{84} = \frac{18}{84} = .214$$

6.

1 3 6: 6 orderings (1 3 6; 1 6 3; 3 1 6; 3 6 1; 6 1 3; 6 3 1)

1 4 5: 6 orderings (1 4 5; 1 5 4; 4 1 5; 4 5 1; 5 1 4; 5 4 1)

2 2 6: 3 orderings (2 2 6; 2 6 2; 6 2 2)

2 3 5: 6 orderings (2 3 5; 2 5 3; 3 2 5; 3 5 2; 5 2 3; 5 3 2)

2 4 4: 3 orderings (2 4 4; 4 2 4; 4 4 2)

3 3 4: 3 orderings (3 3 4; 3 4 3; 4 3 3)

Total: 27 different ways

Probability of getting 10 $= \frac{27}{6^3} = \frac{27}{216} = .125$

UNIT 4

Random Variable Distributions

DISCRETE RANDOM VARIABLES: DICE

If you toss a die, there are six possible outcomes, each equally likely. Make a table of the probabilities:

Possible values	Probabilities
1	1/6
2	1/6
3	1/6
4	1/6
5	1/6
6	1/6
TOTAL	6/6

It would be nice to have a shorter way of writing "the number that appears on the die," so we will call it X. The quantity X is called a *random variable*, because its value depends on the result of a random experiment. More specifically, X is a *discrete random variable*, because it is possible to list its possible values. (By contrast, a *continuous random variable* can take on any real number value in a certain range.) In this case, there are only six possible values. We can rewrite our table in terms of the random variable X. To save us a bit of writing, we can use i to represent one of the specific values. Then we can say:

i	$\Pr(X = i)$	$\Pr(X = i)$
1	1/6	0.1667
2	1/6	0.1667
3	1/6	0.1667
4	1/6	0.1667
5	1/6	0.1667
6	1/6	0.1667
TOTAL	6/6	1.0

Sometimes it is easier to express probabilities as fractions; other times it is best to use decimals. We have done it both ways here so you can compare them.

There is one unbreakable law of discrete random variables: the sum of the probabilities for all possible values must equal 1.

The probability that X equals i is sometimes abbreviated $f(i)$, where f is called the *probability function* for the random variable X.

Random variables arise in many different contexts, including games of chance, sampling, experimental measurements, and natural phenomena. As we did with probability, we will first consider random variables associated with simple random experiments, such as dice and cards. In those cases it is possible to derive the probability for each possible value from a knowledge of the nature of the outcomes of the process.

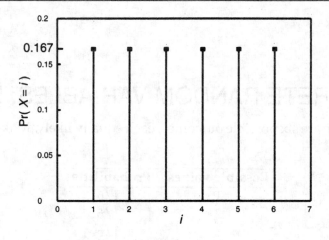

Figure 4.1

Figure 4.1 shows the probabilities for X. Note that the graph is drawn showing spikes at each possible value; the probability is 0 everywhere else. This random variable has the special property that the probability is the same for each possible value.

Things become more interesting if we toss two dice. Let T_2 be the total of the two numbers that appear. We can figure out the probabilities for T_2 by listing all 36 outcomes and counting (see Figure 4.2). To find the probability that T_2 is 7, first count the number of outcomes with a total of 7. There are six of them:

$$(6, 1), (5, 2), (4, 3), (3, 4), (2, 5), (1, 6)$$

$$(1, 1) \quad (1, 2) \quad (1, 3) \quad (1, 4) \quad (1, 5) \quad (1, 6)$$
$$(2, 1) \quad (2, 2) \quad (2, 3) \quad (2, 4) \quad (2, 5) \quad (2, 6)$$
$$(3, 1) \quad (3, 2) \quad (3, 3) \quad (3, 4) \quad (3, 5) \quad (3, 6)$$
$$(4, 1) \quad (4, 2) \quad (4, 3) \quad (4, 4) \quad (4, 5) \quad (4, 6)$$
$$(5, 1) \quad (5, 2) \quad (5, 3) \quad (5, 4) \quad (5, 5) \quad (5, 6)$$
$$(6, 1) \quad (6, 2) \quad (6, 3) \quad (6, 4) \quad (6, 5) \quad (6, 6)$$

Figure 4.2

Second, divide by the total number of outcomes (36). Therefore, the probability of 7 is 6/36 = 1/6. Likewise, we can count the number of outcomes for each possible value from 2 up to 12 and construct a table:

i	# of outcomes	$\Pr(T_2 = i)$	$\Pr(T_2 = i)$
2	1	1/36	0.0278
3	2	2/36	0.0556
4	3	3/36	0.0833
5	4	4/36	0.1111
6	5	5/36	0.1389
7	6	6/36	0.1667
8	5	5/36	0.1389
9	4	4/36	0.1111
10	3	3/36	0.0833
11	2	2/36	0.0556
12	1	1/36	0.0278
TOTAL	36	36/36	1.0

Figure 4.3 shows these probabilities; note how they are highest in the middle. Also note the symmetry.

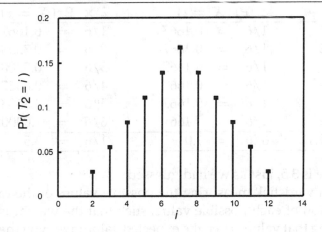

Figure 4.3

EXPECTED VALUE

If a malicious stranger forces you to gamble on the toss of a single die, you have no hope of predicting the result by any means other than a lucky guess. However, suppose the stranger will toss two dice, requiring you to guess close to the average value that appears. To make it concrete, suppose the stranger will allow you to be off by 0.5 in either direction. Now you have a much better chance of being able to guess. Don't guess 1, since the two dice will average 1 only if they both come up 1—and the probability of that is only 1/36. Likewise, don't guess 6, since the probability of both being 6 is only 1/36. You should guess 3.5, since the average will be 3.5 if you get any one of these six outcomes:

$$(1, 6), (2, 5), (3, 4), (4, 3), (5, 2), (6, 1)$$

If the stranger is not mathematically sophisticated, you might be able to convince him to let you toss 50 dice and guess the average. The chance of getting an extreme value (1 or 6) is even less likely, since that would require that all 50 dice took on that extreme value. Most likely you will get some low values and some high values; when you average them out, the result will tend to converge near the middle.

These intuitive results suggest that it is important to look at the average value of the random variable that appears if you repeat the random process many times. This quantity is called the *expected value* (or *expectation*) of the random variable, and is designated by a capital E, as in $E(X)$ for the expected value of the random variable X. Another term for the same idea is *mean*, and the mean of X is designated by the Greek letter mu (μ), pronounced "mew."

The expected value of a discrete random variable can be calculated by finding a weighted average of the possible values. We don't want to take a simple average, since some of the possible values have higher probabilities of occurrence—we want to give those values greater weight. The procedure is to create a table where one column contains the possible values (designated by i), one column shows $\Pr(X = i)$, and one shows $i \times \Pr(X = i)$. Add together the values in the third column, and the result is the expected value. Here are the calculations for $E(X)$:

i	$\Pr(X = i)$		$i \times \Pr(X = i)$	
1	1/6 =	0.1667	1/6 =	0.1667
2	1/6 =	0.1667	2/6 =	0.3333
3	1/6 =	0.1667	3/6 =	0.5000
4	1/6 =	0.1667	4/6 =	0.6667
5	1/6 =	0.1667	5/6 =	0.8333
6	1/6 =	0.1667	6/6 =	1.0000
TOTAL	6/6 =	1.0	21/6 =	3.5

The expected value is 3.5, just as we had guessed.

Imagine you had a yardstick measuring the possible values of the random variable. Put a weight at the location of each possible value, such that the weight is proportional to the probability of attaining that value. Then the expected value gives you the location of the place where the yardstick would balance. Note that the expected value does not itself have to be one of the possible values. You could never actually toss 3.5 on a single roll of the die.

EXAMPLE 1

Calculate the expected value of T_2, the total of the numbers that appear on two dice:

i	# of outcomes	$\Pr(T_2 = i)$			$i \times \Pr(T_2 = i)$		
2	1	1/36	=	0.0278	2/36	=	0.0556
3	2	2/36	=	0.0556	6/36	=	0.1667
4	3	3/36	=	0.0833	12/36	=	0.3333
5	4	4/36	=	0.1111	20/36	=	0.5556
6	5	5/36	=	0.1389	30/36	=	0.8333
7	6	6/36	=	0.1667	42/36	=	1.1667
8	5	5/36	=	0.1389	40/36	=	1.1111
9	4	4/36	=	0.1111	36/36	=	1.0000
10	3	3/36	=	0.0833	30/36	=	0.8333
11	2	2/36	=	0.0556	22/36	=	0.6111
12	1	1/36	=	0.0278	12/36	=	0.3333
TOTAL	36	36/36	=	1.0	252/36	=	7.0

Using the same method as before, we find that $E(T_2) = 7$.

THE EXPECTED VALUE OF THE AVERAGE OF INDEPENDENT, IDENTICALLY DISTRIBUTED RANDOM VARIABLES

Let X_1 be the number on the first die, and X_2 be the number on the second die. We can tell that $E(X_1) = 3.5$ and $E(X_2) = 3.5$, since the probabilities for these two random variables are identical. (Note: This does not mean that the actual value of X_1 must equal the actual value of X_2; only their expected values are equal.) By definition:

$$T_2 = X_1 + X_2$$

We found that:

$$E(T_2) = E(X_1 + X_2) = E(X_1) + E(X_2)$$

Dare we hope to have found a general rule? It turns out that this rule does work for any two random variables X and Y:

> If X and Y are any two random variables, then
> $$E(X + Y) = E(X) + E(Y)$$

Now we investigate what happens if a constant multiplies a random variable.

EXAMPLE 2

Suppose you have the good fortune of being able to play a game where you toss one die and then will be paid \$100 multiplied by the number that appears on the die. What is the expected value of your winnings W?

i	$\Pr(W = i)$			$i \times \Pr(W = i)$		
100	1/6	=	0.1667	100/6	=	16.67
200	1/6	=	0.1667	200/6	=	33.33
300	1/6	=	0.1667	300/6	=	50.00
400	1/6	=	0.1667	400/6	=	66.67
500	1/6	=	0.1667	500/6	=	83.33
600	1/6	=	0.1667	600/6	=	100.00
TOTAL	6/6	=	1.0	2100/6	=	350.00

Therefore, $E(W) = 350$. Suppose that you were charged \$350 each time you played this game. This result suggests that if you played the game many times, you would tend to just break even.

Note that W can be found from this formula:

$$W = 100X$$

Also note:

$$E(W) = E(100X) = 100E(X)$$

This suggests another general rule:

> If c is any constant, and X is any random variable, then:
> $$E(cX) = cE(X)$$

We have previously found $E(X) = 3.5$ and $E(T_2) = 7$. Let A_2 be the average of X_1 and X_2. To find $E(A_2)$, note that $A_2 = T_2/2$. Therefore:

$$E(A_2) = E(\frac{1}{2}T_2) = \frac{1}{2}E(T_2) = \frac{1}{2} \times 7 = 3.5$$

In general, suppose $X_1, X_2, ..., X_n$ are n random variables that all have identical probabilities. Let μ represent the expected value of each of them: $E(X_1) = \mu$, $E(X_2) = \mu$, and so on.

If you toss n dice, and let X_j be the number that appears on dice j, then it qualifies for application of this rule. Let T_n be the sum of all of these random variables:

$$T_n = X_1 + X_2 + ... + X_n = \sum_{j=1}^{n} X_j$$

Then:

$$
\begin{aligned}
E(T_n) &= E(X_1 + X_2 + \ldots + X_n) \\
&= E(X_1) + E(X_2) + \ldots + E(X_n) \\
&= \mu + \mu + \ldots + \mu \\
&= \sum_{j=1}^{n} \mu = n\mu
\end{aligned}
$$

Now let A_n be the average of these n random variables:

$$
A_n = \frac{T_n}{n}
$$

To find $E(A_n)$:

$$
E(A_n) = E\left(\frac{T_n}{n}\right) = \frac{1}{n}E(T_n) = \frac{n\mu}{n} = \mu
$$

Therefore, the expected value of the average A_n is the same as the expected value of the original random variable. For the dice example, $E(A_n) = \mu = 3.5$. This fact seems rather obvious and unexciting, but we will soon come across a very striking property of A_n.

VARIANCE

Now we need to investigate the degree of uncertainty in a random variable. Consider these six random variables:

V: 96.5% chance of being 0; 3.5% chance of being 100
U: 50% chance of being 1; 50% chance of being 6
X: the number that appears when one die is tossed
A_2: the average number that appears when two dice are tossed
A_3: the average number that appears when three dice are tossed
A_7: the average number that appears when seven dice are tossed

All of these random variables have one thing in common: their expected value is 3.5. Find $E(V)$ and $E(U)$ from these tables:

i	$\Pr(V = i)$	$i \times \Pr(V = i)$
0	0.965	0
100	0.035	3.5
TOTAL	1.0	$E(V) = 3.5$

i	$\Pr(U = i)$	$i \times \Pr(U = i)$
1	0.5	.5
6	0.5	3.0
TOTAL	1.0	$E(U) = 3.5$

Other than equal expected values, the behavior of our six random variables is quite different. There is a lot of risk associated with random variable V; the risk decreases as we move

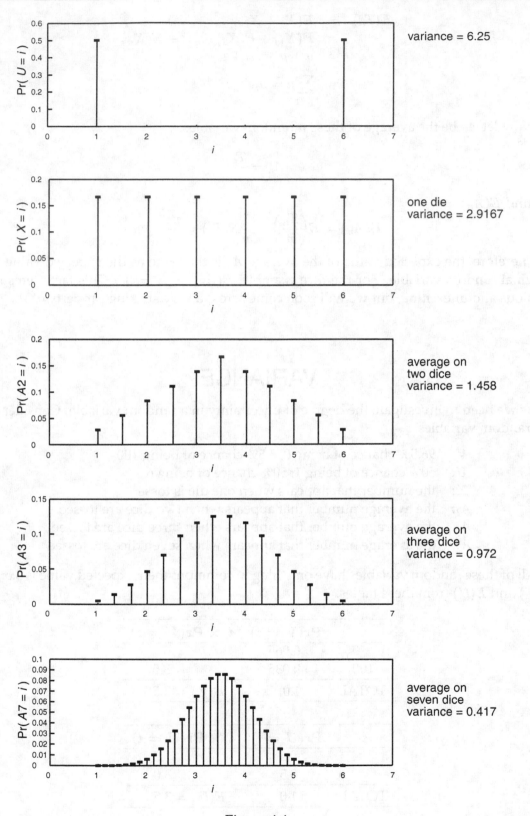

Figure 4.4

down the list. Figure 4.4 shows the probabilities for five of these random variables (not including V because it won't fit on the scale). In each case, the sum of the heights of the spikes is 1. At the top, the probability is spread out wide; further down the page, the probability becomes more concentrated into the middle.

We need a way of measuring the uncertainty of a random variable. The concept is similar to the concept in Unit 2, where the variance was used to measure the dispersion of a list of numbers. Recall that we found the variance of a list by looking at the difference between each number from the mean. We could not simply average these differences because then the positives and negatives would cancel out, so we squared the differences and then averaged. We can use the same procedure for a random variable, except we have to weight each squared distance by the probability of that value occurring.

Here are the calculations for the variance of X (which is abbreviated Var(X), or σ^2 (sigma-squared)):

i	$\Pr(X = i)$	$i - E(X)$	$(i - E(X))^2$	$(i - E(X))^2 \times \Pr(X = i)$
1	1/6	−2.5	6.25	1.0417
2	1/6	−1.5	2.25	0.3750
3	1/6	−0.5	0.25	0.0417
4	1/6	0.5	0.25	0.0417
5	1/6	1.5	2.25	0.3750
6	1/6	2.5	6.25	1.0417
TOTAL	6/6	0	17.50	2.9167

The steps are:

1. Calculate $i - E(X) = i - 3.5$ for each possible value.

2. Square these possible values.

3. Multiply these squares by the probability (see the last column).

4. Add together the final column; the result is Var(X).

In this case, Var(X) = 2.9167.

There is a slightly shorter way for finding Var(X). Create a column with the values of i^2; multiply each of these squares by the probability; then add the sum of these values. The result is the expected value of the square of X, written $E(X^2)$:

i	$\Pr(X = i)$		i^2	$i^2 \times \Pr(X = i)$	
1	1/6	0.1667	1	1/6	0.1667
2	1/6	0.1667	4	4/6	0.6667
3	1/6	0.1667	9	9/6	1.5000
4	1/6	0.1667	16	16/6	2.6667
5	1/6	0.1667	25	25/6	4.1667
6	1/6	0.1667	36	36/6	6.0000
TOTAL	6/6	1		91/6	15.1667

Therefore, $E(X^2) = 15.1667$. Once this is known, find the variance from this formula:

$$\text{Var}(X) = E(X^2) - [E(X)]^2$$

For the case of the single die:

$$\text{Var}(X) = 15.1667 - 3.5^2 = 2.9167$$

This is the same answer we found from the other method.

Here is another example, showing the calculation for $\text{Var}(T_2)$:

i	# of outcomes	$\text{Pr}(T_2 = i)$	$i^2 \times \text{Pr}(T_2 = i)$	
2	1	1/36	$4 \times 1/36$	0.1111
3	2	2/36	$9 \times 2/36$	0.5000
4	3	3/36	$16 \times 3/36$	1.3333
5	4	4/36	$25 \times 4/36$	2.7778
6	5	5/36	$36 \times 5/36$	5.0000
7	6	6/36	$49 \times 6/36$	8.1667
8	5	5/36	$64 \times 5/36$	8.8889
9	4	4/36	$81 \times 4/36$	9.0000
10	3	3/36	$100 \times 3/36$	8.3333
11	2	2/36	$121 \times 2/36$	6.7222
12	1	1/36	$144 \times 1/36$	4.0000
TOTAL	36	36/36	1974/36	$E(T_2^2) = 54.8333$

We find $\text{Var}(T_2)$ is $54.8333 - 7^2 = 5.83333$, which happens to be $2.91667 + 2.91667$. Since $\text{Var}(T_2) = \text{Var}(X_1 + X_2) = \text{Var}(X_1) + \text{Var}(X_2)$ we're tempted to state a general rule about the variance of a sum. This rule does work, but only under the right condition:

> If X and Y are any two *independent* random variables, then
> $$\text{Var}(X + Y) = \text{Var}(X) + \text{Var}(Y)$$

Two random variables are independent if knowledge of the value for one of them provides no help in predicting the value of the other one. (Note that this definition is similar to the definition of independent events.) Since the number on one die is not affected by the number on the other die, these two random variables qualify as being independent. (See **covariance** in Part II for information on how to measure independence, and how to find $\text{Var}(X + Y)$ if X and Y are not independent.)

Here are the calculations for $\text{Var}(U)$ and $\text{Var}(V)$:

i	$\text{Pr}(V = i)$	i^2	$i^2 \times \text{Pr}(V = i)$
0	0.965	0	0
100	0.035	10,000	350
TOTAL	1		$E(V^2) = 350$

Therefore: $\text{Var}(V) = E(V^2) - [E(V)]^2 = 350 - 3.5^2 = 337.75$.

i	$\text{Pr}(U = i)$	i^2	$i^2 \times \text{Pr}(U = i)$
1	0.5	1	.5
6	0.5	36	18.0
TOTAL	1		$E(U^2) = 18.5$

Therefore: $\text{Var}(U) = E(U^2) - [E(U)]^2 = 18.5 - 3.5^2 = 6.25$.

Figure 4.4 shows how the variance becomes smaller as the probabilities become concentrated closer to the middle.

Two other properties to note follow:

- The variance will be zero if and only if X is a constant—that is, if it is not really a random variable at all.

- The variance is never negative.

The square root of the variance is called the *standard deviation*, represented by σ (sigma). We can calculate the standard deviation for each of our examples:

Random variable	Variance (σ^2)	Standard deviation (σ)
V	337.75	18.378
U	6.25	2.500
X	2.9167	1.708
A_2	1.458	1.207
A_3	0.972	0.986
A_7	0.417	0.646

We might be tempted to say that $\text{Var}(cX) = c\text{Var}(X)$, but matters are not quite that simple. Because of the way that calculating a variance requires squaring the values, you must also square a constant if it is pulled outside of a variance:

> If c is any constant, and X is any random variable, then:
> $$\text{Var}(cX) = c^2\text{Var}(X)$$

THE LAW OF LARGE NUMBERS

We can now put together a formula for the variance of the average A_n. Let X_1, X_2, ..., X_n be n independent random variables that all have identical probabilities. Let μ represent the expected value of each of them: $E(X_1) = \mu$, $E(X_2) = \mu$, and so on. Let T_n be the sum of all of these random variables:

$$T_n = X_1 + X_2 + \ldots + X_n = \sum_{j=1}^{n} X_j$$

Apply the formula for the variance of a sum:

$$\begin{aligned}
\text{Var}(T_n) &= \text{Var}(X_1 + X_2 + \ldots + X_n) \\
&= \text{Var}(X_1) + \text{Var}(X_2) + \ldots + \text{Var}(X_n) \\
&= \sigma^2 + \sigma^2 + \ldots + \sigma^2 \\
&= \sum_{j=1}^{n} \sigma^2 \\
&= n\sigma^2
\end{aligned}$$

Let A_n be the average of these n random variables:

$$A_n = \frac{T_n}{n}$$

To find $\text{Var}(A_n)$:

$$\text{Var}(A_n) = \text{Var}\left(\frac{T_n}{n}\right) = \frac{1}{n^2}\text{Var}(T_n) = \frac{n\sigma^2}{n^2} = \frac{\sigma^2}{n}$$

The average is also symbolized by $\bar{x}$, so we can write:

$$\text{Var}(\bar{x}) = \frac{\sigma^2}{n}$$

Notice an interesting fact that turns out to be of profound importance: the variance of the average of a group of identically distributed independent random variables is not the same as the variance of the individual random variables—it is less. In fact, it is potentially much less, since σ^2/n will become smaller and smaller as n becomes bigger and bigger. We intuitively felt that if you tossed many dice, the average would tend to converge to the expected value, 3.5. Now we have mathematical justification for asserting that fact. In general, as the number of trials approaches infinity, the average A_n must converge to the expected value μ. This proposition is known as the *law of large numbers*. If this proposition weren't true, we would have to give up in despair at this point—there would be no hope for statistical inference. However, since it is true, we can have confidence that a large enough sample will be representative of the population—provided that the sample is selected randomly. (Unit 5 states this idea more precisely.)

For our examples:

$$\text{Var}(A_3) = \frac{2.91667}{3} = 0.972$$

$$\text{Var}(A_7) = \frac{2.91667}{7} = 0.417$$

Figure 4.4 shows how the probabilities converge to the middle as n becomes larger and the variance becomes smaller.

SPECIAL DISCRETE RANDOM VARIABLE DISTRIBUTIONS

Several types of random variable distributions have so many important applications that they have been given special names. (See the entry for each distribution in Part II for more specific information.)

The *binomial distribution* applies to a situation where you conduct n trials of some kind of process, where the probability of success on each trial is p. If your chance of success is .34, and you conduct 100 trials, then you would expect 34 successes. In general, the expected value of a binomial random variable is np and its variance is $np(1-p)$. A generalization of the binomial distribution is the *multinomial distribution*. The *Poisson distribution* can be used as an approximation for the binomial distribution in some cases. It describes certain types of random processes, such as the number of phone calls that arrive at an office in a particular unit of time.

The *hypergeometric distribution* applies when you are sampling without replacement. If you have $N = 1,000$ marbles in a jar ($M = 600$ red and $N - M = 400$ blue), and you randomly select $n = 10$ marbles, letting X represent the number of red marbles you select, then X is a random variable with the hypergeometric distribution with parameters $N = 1,000$, $M = 600$, and $n = 10$. Its expected value is

$$\left(\frac{M}{N} \right) n = \left(\frac{600}{1,000} \right) \times 10 = 6$$

For other examples of discrete random variables, see **geometric distribution** and **negative binomial distribution** in Part II.

CONTINUOUS RANDOM VARIABLES

Toss 1,000 dice, and calculate the probability that the average of these 1,000 numbers ($A_{1,000}$) will be between 3.5 and 3.6 (see Figure 4.5). We would have to add up all of the probabilities: $\Pr(A_{1,000} = 3.500)$; $\Pr(A_{1,000} = 3.501)$; $\Pr(A_{1,000} = 3.502)$; and so on. There are too many lines to show them all in the figure. If you could fill them all in, the area under the curve would appear solid black—which raises an interesting question: perhaps it would be easier simply to find the area under the probability curve?

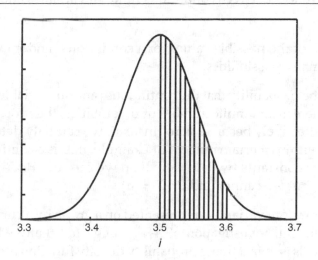

Figure 4.5

Our random variable $A_{1,000}$ is still a discrete random variable, even though it has many possible values that are all close together: 3.501; 3.502; 3.503; and so on. There are other random variables that can take on any real number value within a certain range; perhaps 3.5000001, 3.500002, and an infinite number of other fractional values between these two. Such a random variable is called a *continuous random variable*. These variables often arise in practice; for example, if you are to measure time, distance, area, volume, mass, energy, voltage, temperature, or pressure, the result can vary continuously over all real number values. A discrete random variable, such as $A_{1,000}$, which has many possible values that are close together, can

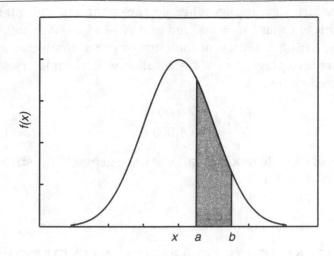

Figure 4.6: If $f(x)$ represents the probability density function for a continuous random variable X, then the area under the curve between a and b gives the probability that X will be between a and b.

often be approximated as if it were a continuous random variable. "Why do that?" you might ask. The answer is that continuous random variables often are easier to deal with.

Here are the differences between continuous random variables and discrete random variables:

1. We can't list all of the possible values for a continuous random variable, because there are infinitely many possibilities.

2. We can't find the probability that the continuous random variable will equal a particular value. Since there is an infinite number of possibilities, the probability of hitting one of them exactly effectively becomes zero. Instead, we can only determine the probability that it will be within a certain range. For example, if X is a continous random variable, and a and b are constants (with $b > a$), then we can find $\Pr(X < a)$ or $\Pr(X > a)$ or $\Pr(a < X < b)$, but we cannot find $\Pr(X = a)$.

3. A continuous random variable is represented on a graph by a curve called the *probability density function*. Our investigation of $A_{1,000}$ suggested that we look at the area under such a curve. This is in fact how a probability density function is defined: the area under the curve between $X = a$ and $X = b$ is the probability that the random variable X will be between a and b (see Figure 4.6).

    Of course, this only helps if there is a convenient way to find the area under the curve. Sometimes it is possible to use calculus to find a formula for the area under the curve if you know the formula for the curve itself (see **integral** in Part II). Sadly, it often happens in statistics that the curves we are interested in are such that it is impossible to find a simple formula for the area. In that case, we have two options: (1) look up the result in a table (some standard statistical tables are included at the back of the book), or (2) use a built-in function in your calculator or a computer program such as Microsoft Excel.

4. There is one unbreakable law of continuous random variables: the total area under the density function must be 1.

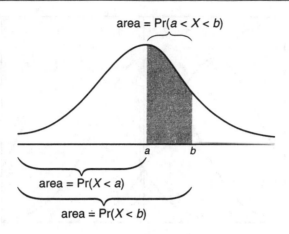

area = $\Pr(a < X < b)$

area = $\Pr(X < a)$

area = $\Pr(X < b)$

Figure 4.7

5. $\Pr(X > a) = 1 - \Pr(X < a)$

6. $\Pr(a < X < b) = \Pr(X < b) - \Pr(X < a)$
   (assuming $a < b$; see Figure 4.7).

Fortunately, continuous random variables do have some properties that are the same as we found for discrete random variables. They have expected values that mean the same thing (the average that would appear if you observed the random variable many times) even though they are calculated differently. The variance measures the degree to which the probabilities are spread out; the next section illustrates some examples.

THE NORMAL DISTRIBUTION

As the number of dice increases, we found that the probability function for the average takes on the shape of a nice bell-shaped curve—truly a thing of beauty. This curve applies to many other situations as well, so it has become known as the *normal curve*. Technically, it is the density function for a continuous random variable that has the *normal distribution*. Some of the situations where it applies include:

- The distribution of a quantity where extreme values are less likely than central values (heights, weights, test scores, etc.) typically has a normal distribution.

- The random errors associated with the measurement of a physical quantity often have a normal distribution.

- When a random process is repeated many times, then the normal distribution can often be used as a convenient approximation for other kinds of distributions. We have already seen how this happens with dice (see **Central Limit Theorem** in Part II for more information).

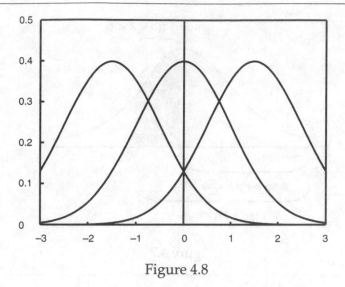

Figure 4.8

The density function curve for the normal random variable is given by:

$$f(x) = \frac{1}{\sigma\sqrt{2\pi}}e^{-\frac{1}{2}\left(\frac{x-\mu}{\sigma}\right)^2}$$

You need to know the value of the mean μ and the standard deviation σ in order to use this formula. (Fortunately, in most problems you either are given those values or else there are standard ways to calculate them.) The letter e represents a special number whose value is about 2.71828 (see **e** in Part II). You will not likely need to calculate the height of the density function very often; it is the area you will need to know.

The value of μ determines the location of the middle, or peak of the distribution. Figure 4.8 shows several different normal distributions with different values of μ. All of these have the same shape, because they all have the same value of σ.

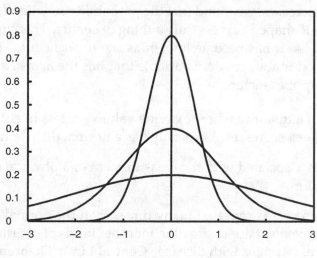

Figure 4.9

Figure 4.9 shows what happens as the value of σ changes while μ stays the same. A smaller value of σ means the probabilities are concentrated closer to the middle, which leads to a higher, narrower curve. Remember that in all cases the total area under the curve must be 1.

   Unfortunately, there is no simple formula for the area under the normal distribution curve, so we have to use a computer or a table. It would be hopeless if we needed a table for every possible value of μ and σ; fortunately, it turns out that we only need a table for one situation: $\mu = 0, \sigma = 1$, which is called the *standard normal distribution*. If X has a normal distribution with mean μ and standard deviation σ, then we can create a new random variable Z from this formula:

$$Z = \frac{X - \mu}{\sigma}$$

The variable Z will have a standard normal distribution. To find the probability that X will be less than a:

$$\Pr(X < a) = \Pr\left(\frac{X - \mu}{\sigma} < \frac{a - \mu}{\sigma}\right) = \Pr\left(Z < \frac{a - \mu}{\sigma}\right)$$

We can look up the probability that Z is less than any particular value in the standard normal random variable table at the back of the book.

EXAMPLE 3

We still suspensefully await the solution to $\Pr(3.5 < A_{1,000} < 3.6)$. Since the variance of the number on one die is $\sigma^2 = 2.91667$, we know

$$\text{Var}(A_{1,000}) = \frac{\sigma^2}{n} = \frac{2.91667}{1,000} = 0.002917$$

The standard deviation of $A_{1,000}$ is $\sqrt{0.002917} = 0.054$, so we will approximate $A_{1,000}$ by a normal random variable A' with mean 3.5 and standard deviation 0.054. Then:

$$\begin{aligned}
\Pr(3.5 < A' < 3.6) &= \Pr\left(\tfrac{3.5-3.5}{.054} < \tfrac{A'-3.5}{.054} < \tfrac{3.6-3.5}{.054}\right) \\
&= \Pr(0 < Z < 1.85) \\
&= \Pr(Z < 1.85) - \Pr(Z < 0)
\end{aligned}$$

Look in the standard normal table (page 205): $\Pr(Z < 1.85) = 0.9678$; $\Pr(Z < 0) = 0.5$; so our solution is $0.9678 - 0.5 = 0.4678$ (see Figure 4.10).

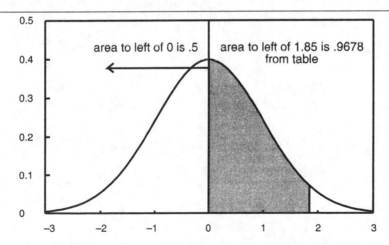

Figure 4.10: $\Pr(0 < Z < 1.85) = 0.9678 - 0.5000 = 0.4678$

There are several continuous random variable distributions related to the normal that are used extensively in statistics (see **chi-square distribution, t distribution, F distribution** in Part II).

EXERCISES

1. Let T_3 be a random variable equal to the sum of the three numbers that appear when you toss three dice. Determine the probability function for this random variable, and then calculate $E(T_3)$, $E(T_3^2)$, and $\text{Var}(T_3)$.

2. Let T_{75} be a random variable equal to the sum of the 75 numbers that appear if you toss 75 dice. Calculate $E(T_{75})$ and $\text{Var}(T_{75})$. (Do not attempt to find the probability function.)

3. Let A_4 be a random variable equal to the average of the four numbers that appear when you toss four dice. Let A_{75} be a random variable equal to the average of the 75 numbers that appear when you toss 75 dice. Calculate $E(A_4)$, $E(A_{75})$, Var(A_4), and Var(A_{75}).

For Exercises 4 to 8, see the entry in Part II on the **binomial distribution**. Use a computer to do the calculations; see the Appendix.

4. Suppose you are taking a 30-question multiple choice exam. Each question has four choices, and you will select all of your answers by random guessing.
(a) What is the probability you will get two questions correct?
(b) What is the probability you will get at least 5 right?
(c) What is the probability you will get at least 15 right?
(d) What is the probability you will get at least 20 right?
(e) If X is the random variable representing the number of questions you get right, what are $E(X)$ and Var(X)?

5. Suppose you are running an airline that flies planes seating 400 people each. On average, 4% of the people who make reservations fail to show up for their flights. You decide to gamble by taking more than 400 reservations for each flight, but if too many people happen to show up for a particular flight you must pay a steep penalty. (Use a computer to calculate the binomial distribution. See the appendix. You will have to find the answer by trial and error.)

(a) What is the maximum number of reservations you may take if you want the probability of an overflow to be less than 1%?

(b) What is the maximum number of reservations you may take if you want the probability of an overflow to be less than 6%?

(c) What is the maximum number of reservations you may take if you want the probability of an overflow to be less than 15%?

6. Consider the situation in the previous problem. Suppose you decide to take enough reservations so that the probability of an overflow on each particular flight is 6%. Your airline flies 300 flights per week. (All of the flights are overbooked; in other words, the number of reservations for each flight is the number of reservations you found for the answer to Question 5, part b.) Let X represent the number of flights per week where an overflow occurs. What are $E(X)$ and $\text{Var}(X)$? Assume that each flight is independent—in other words, whether or not an overflow occurs on a particular flight has no effect on whether any other flight will have an overflow.

7. Consider an insurance company with 200 customers. From each customer it collects a premium of $20 per year. It must pay $1,000 to a customer who files a claim. The probability that a customer will file a claim in a year is .02. The probability is the same for all customers, and all customers are independent (in other words, whether or not a particular customer files a claim has no effect on whether any other customer will file a claim).

(a) What are the expected value and the variance of the firm's yearly profits? (This imaginary firm has no administrative expenses. Its only revenue comes from premium payments, and its only expenses are claims that must be paid.)

(b) Calculate the probability that the firm's profits will have each of the following values: 4,000; 3,000; 2,000; 1,000; 0; –1,000; –2,000; and –3,000.

---

8. Consider an insurance company with 400 customers. It must pay out $1,000 to a customer who files a claim. The probability that a customer will file a claim in a year is .02. The probability is the same for all customers, and all customers are independent (in other words, whether or not a particular customer files a claim has no effect on whether any other customer will file a claim). What is the smallest premium it may charge while still having a probability greater than 80% that the firm will at least break even in a year (in other words, have a yearly profit greater than or equal to zero)?

9. Suppose your office has been moved to a new location, along with the other offices on your floor. A total of 120 boxes were moved. Of these, 16 belong to you and the other 104 belong to other people. The movers randomly selected 14 out of the 120 boxes to put in your new office. Let $X =$ the number of boxes put in your office that do in fact belong to you. Calculate $\Pr(X = k)$ and $\Pr(X \leq k)$ for $k = 0$ to $k = 14$. (See the entry in Part II on the **hypergeometric distribution.** You may want to use a computer for the calculations; see the appendix.)

10. You have found that the height of adults in a town of 10,000 people is distributed according to a normal distribution with mean 5'8"and standard deviation 5".
(a) How many people are taller than 6'6"?
(b) How many people are between 6'0"and 6'6"?
(c) How many people are between 5'7"and 6'0"?
(d) How many people are between 5'4"and 5'7"?
(e) How many people are shorter than 5'4"?

11. Consider an office with 100 workers. The arrival time for the workers follows a normal distribution with mean 8:56 A.M. and standard deviation 4 minutes.
(a) How many workers will arrive after 9:00 A.M. on average?
(b) How many workers arrive before 8:52 A.M. on average?
(c) How many workers arrive between 8:59 A.M. and 9:00 A.M. on average?

For problems 12, 13, and 14, refer to the Normal Distribution Properties on page 183.

12. The lifetime of a particular electronic circuit has a normal distribution with mean 50,000 hours and standard deviation 8,000 hours. You have selected one of these circuits at random.
(a) What is the probability that your circuit will last less than 30,000 hours?
(b) What is the probability that your circuit will last more than 55,000 hours?
(c) Suppose you buy two circuits. When the first circuit fails you will replace it with the second circuit. Let X be a random variable representing the lifetime of the first circuit plus the lifetime of the second circuit. What is the distribution of X? What is its mean? What is its variance?
(d) What is the probability that X will be greater than 80,000 hours?

13. Suppose you roll 1,000 dice. Let X be a random variable representing the sum of the numbers that appear on the dice.
(a) What will the distribution of X be? What will be its mean and variance?
(b) What is the probability that X will be greater than 4,000?
(c) Let A be a random variable equal to the average of the 1,000 numbers that appear on the dice. What will the distribution of A be? What will be its mean and variance?
(d) What is the probability that A will be between 3.2 and 3.8?

14. Consider a situation where you have a choice between two jobs. Your annual earnings from the office job will have a normal distribution with mean $30,000 and a standard deviation of $2,000. Your annual earnings from a traveling sales job will have a normal distribution with mean $32,000 and standard deviation $12,000. What is the probability that you would earn more from the traveling sales job?

ANSWERS

1.

k	$f(k) = \Pr(T_3 = k)$		$k \times f(k)$	$k^2 \times f(k)$
3	1/216	= 0.0046	0.014	0.042
4	3/216	= 0.0139	0.056	0.222
5	6/216	= 0.0278	0.139	0.694
6	10/216	= 0.0463	0.278	1.667
7	15/216	= 0.0694	0.486	3.403
8	21/216	= 0.0972	0.778	6.222
9	25/216	= 0.1157	1.042	9.375
10	27/216	= 0.1250	1.250	12.500
11	27/216	= 0.1250	1.375	15.125
12	25/216	= 0.1157	1.389	16.667
13	21/216	= 0.0972	1.264	16.431
14	15/216	= 0.0694	0.972	13.611
15	10/216	= 0.0463	0.694	10.417
16	6/216	= 0.0278	0.444	7.111
17	3/216	= 0.0139	0.236	4.014
18	1/216	= 0.0046	0.083	1.500
Total:		1.0000	10.500	119.000

$E(T_3) = 10.5$; $E(T_3^2) = 119$; $\mathrm{Var}(T_3) = 119 - 10.5^2 = 8.75$ (Note: $8.75 = 2.91667 + 2.91667 + 2.91667$, so the variance on the sum of three dice is equal to the sum of the variances on each individual die.)

2. $E(T_{75}) = E(X_1) + E(X_2) + \ldots + E(X_{75})$, where X_i is the number on toss i, and $E(X_i) = 3.5$. Then $E(T_{75}) = 262.5$.
$\mathrm{Var}(T_{75}) = \mathrm{Var}(X_1) + \mathrm{Var}(X_2) + \ldots + \mathrm{Var}(X_{75}) = 218.75$

3. $E(A_4) = E(A_{75}) = 3.5$;
$\mathrm{Var}(A_4) = 2.91667/4 = 0.729$.
$\mathrm{Var}(A_{75}) = \mathrm{Var}(T_{75}/n) = (1/n^2)\mathrm{Var}(T_{75}) = 0.0389$

4. (a) $\Pr(X = 2) = \binom{30}{2}.25^2.75^{28} = .00863$
(b) .902
(c) .00275
(d) less than .00001
(e) $E(X) = np = 7.5$; $\mathrm{Var}(X) = np(1-p) = 5.625$

5. (a) 408 (b) 411 (c) 412

6. Use a binomial distribution with $n = 300$ and $p = .06$ $E(X) = 18$; $\mathrm{Var}(X) = 16.92$

7. Let X be the number of claims and P be the level of profit. X has a binomial distribution with $n = 200$ and $p = .02$. Then $P = 4,000 - 1,000X$; $E(P) = 4,000 - 1,000E(X) = 0$;

$\text{Var}(P) = \text{Var}(4{,}000 - 1{,}000X) = 0 + \text{Var}(-1{,}000X) =$
$(-1{,}000)^2 \text{Var}(X) = 3{,}920{,}000.$
(b)

Claims	Probability	Profits
0	.01759	4,000
1	.07179	3,000
2	.14577	2,000
3	.19635	1,000
4	.19735	0
5	.15788	-1,000
6	.10472	-2,000
7	.05923	-3,000

8. Create a table of binomial distribution probabilities with $n = 400$ and $p = .02$. You will find there is a probability of .818 of having 10 or fewer claims. It can charge a premium of $25 and still break even with 10 claims.

9. Use the hypergeometric distribution with $N = 120, M = 16, n = 14$.

k	Pr($X \leq k$)	Pr($X = k$)
0	.1188	.1188
1	.4112	.2924
2	.7210	.3099
3	.9076	.1866
4	.9786	.0710
5	.9965	.0179
6	.9996	.0031
7	1.0000	.0004
8	1.0000	.0000 [2.9×10^{-05}]
9	1.0000	.0000 [1.6×10^{-06}]
10	1.0000	.0000 [5.5×10^{-08}]
11	1.0000	.0000 [1.2×10^{-09}]
12	1.0000	.0000 [1.5×10^{-11}]
13	1.0000	.0000 [8.7×10^{-14}]
14	1.0000	.0000 [1.8×10^{-16}]

10. $\mu = 68$; $\sigma = 5$
(a) $\Pr(X > 78) = \Pr[(X - 68)/5 > (78 - 68)/5] = \Pr(Z > 2) =$
$1 - \Pr(Z < 2) = 1 - .9772 = .0228$
Out of 10,000 people, we can expect 228 people will have a height in this range.
(b) 1,891 (c) 3,674
(d) 2,088
(e) 2,119

11. Measure time in minutes with 9:00 represented as zero; then $\mu = -4$ and $\sigma = 4$.
(a) $\Pr(X > 0) = \Pr[\frac{(X-(-4))}{4} > \frac{(0-(-4))}{4}] = \Pr(Z > 1) = 1 - \Pr(Z < 1) = 1 - .8413 = .1587$

Out of 100 workers, expect an average of 15.9 workers to arrive after 9:00 A.M.

(b) 15.9

(c) 6.8

12. (a) .0062

(b) .2660

(c) X will have a normal distribution (because of the addition property of normal random variables).

$E(X) = 50,000 + 50,000 = 100,000$

$\text{Var}(X) = 8,000^2 + 8,000^2 = 128,000,000$

standard deviation = 11,313.708

(d) $\Pr(X > 80,000) = .9614$

13. Because of the addition properties for expectation and variance, $E(X) = 3,500$ and $\text{Var}(X) = \text{Var}(X_1) + \text{Var}(X_2) + \ldots + \text{Var}(X_{1000})$ where $\text{Var}(X_i)$ is the variance on dice i, which is 2.916667; so $\text{Var}(X) = 2,916.667$ and the standard deviation is 54.006. Because of the central limit theorem, X will have a normal distribution.

(b) $\Pr(X > 4,000) = \Pr[Z > \frac{(4,000-3,500)}{54.006}] = \Pr(Z > 9.26)$

The table does not include this value, but clearly it must be very close to zero.

(c) A will have a normal distribution. $A = X/1,000$, so $E(A) = E(X)/1,000 = 3.5$ and $\text{Var}(A) = (\frac{1}{1,000})^2 \text{Var}(X) = 0.00292; \sigma = 0.054$

$\Pr(3.2 < A < 3.8) = \Pr(\frac{-0.3}{.054} < Z < \frac{0.3}{0.054}) = \Pr(Z < 5.56) - \Pr(Z < -5.56)$ The table does not include these values, but $\Pr(Z < 5.56)$ is very close to 1, and $\Pr(Z < -5.56)$ is very close to zero, so $\Pr(3.2 < A < 3.8)$ is very close to 1.

14. Let X = earnings from office job; Y = earnings from traveling job. $\Pr(Y > X) = \Pr[(Y - X) > 0]$. Let $D = Y - X$. D will have a normal distribution. Then $E(D) = E(Y) - E(X) = 2,000;$

$\text{Var}(D) = \text{Var}[Y + (-1)X] = \text{Var}(Y) + (-1)^2 \text{Var}(X)$

$= 12,000^2 + 2,000^2 = 148$ million;

$\sigma = 12,165.53;$

$\Pr(D > 0) = \Pr[\frac{(D-2,000)}{12,165.53} > \frac{(0-2,000)}{12,165.53}]$

$= \Pr(Z > -0.16) = \Pr(Z < 0.16) - 0.5636$

UNIT 5

Polls, Sampling, and Confidence Intervals

Like magic, polls are often able to predict the attitudes of a large group by polling only a small sample of that group. For example, news organizations are often able to predict the results of an election by reporting the result of a poll of only a few hundred voters. At first, you should find it hard to believe that the opinions of a small random sample would match the opinions of a very large population. This unit will show you how much you can trust polls by turning the question around: it would be very unlikely for the opinions of the sample *not* to match the opinions of the population. (However, there are some important disclaimers that are discussed at the end of the unit.)

DRAWING CARDS

Shuffle a standard deck of 52 cards, and deal out 10 cards at random. Count the number of hearts that appear among those 10 cards (let that number equal X). What are the probabilities that X will take on each of the values from 0 up to 10?

We should be able to make some intuitive predictions. You expect that the chances of drawing 10 hearts is very small. On the other hand, the chance of drawing 0 hearts also seems small (although drawing 0 hearts is more likely than drawing 10 hearts). What value do you expect is most likely for the number of hearts? Since one-quarter of the cards in the population (deck) are hearts, you should expect that most likely the number of hearts in the sample will be about one-quarter of the total. Combining these intuitive predictions, we expect a graph of the probabilities for the number of hearts would appear as shown in Figure 5.1. The graph

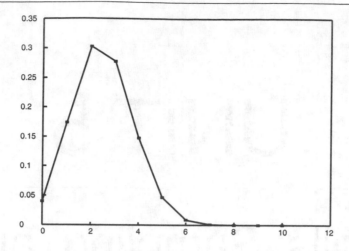

Figure 5.1: Hypergeometric distribution: $N = 52, n = 10, M = 13$

shows the probability that $X = i$, where i runs from 0 to 10. Note that the highest probability occurs where $i = 2$. One-quarter of 10 is 2.5, but the number of hearts must be a whole number, so it is most likely to be about 2 or 3. The probabilities fall steeply the further away from 2 you go in either direction. The total of all of these probabilities is equal to 1, as must be the case for a discrete random variable. The exact formula for these probabilities comes from the *hypergeometric distribution*, where N is the number of items in the population (52), n is the number of items in the sample (10), and M is the number of items of interest in the population (13 in this case, since there are 13 hearts in the deck).

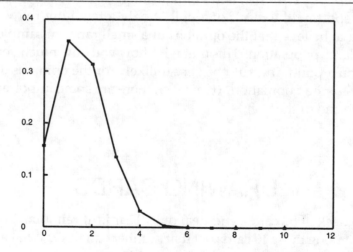

Figure 5.2: Hypergeometric distribution: $N = 52, n = 10, M = 8$

Suppose instead of being interested in hearts, we were interested in royal cards. There are 8 royal cards (kings and queens) in the deck. If we draw 10 cards at random, we know there is 0 probability of getting 9 or 10 royal cards; the probabilities of getting 6, 7, or 8 royal cards are nonzero but very close to zero. The fraction of royal cards in the deck is $8/52 = .154$, so we would expect the probabilities for the 10-card sample to peak at a whole number near

$10 \times .154 = 1.54$. The actual probabilities, shown in Figure 5.2, match our predictions. Most likely we will draw one royal card in our sample.

If we are interested in the number of non-hearts we draw, the probabilities would appear as shown in Figure 5.3, which peaks at 8.

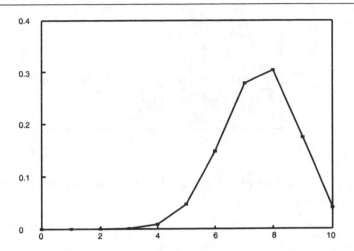

Figure 5.3: Hypergeometric distribution: $N = 52, n = 10, M = 39$

If we are interested in the number of red cards we draw, the probabilities are shown in Figure 5.4. These probabilities are symmetric about 5.

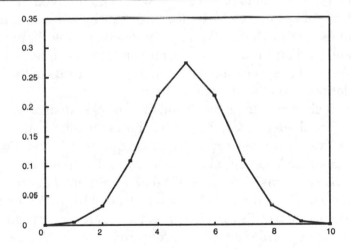

Figure 5.4: Hypergeometric distribution: $N = 52, n = 10, M = 26$

SAMPLING WITH AND WITHOUT REPLACEMENT

When we draw a card from the deck, we keep the card in our hand, rather than returning it to the deck before drawing the next card. This sampling procedure is called *sampling without*

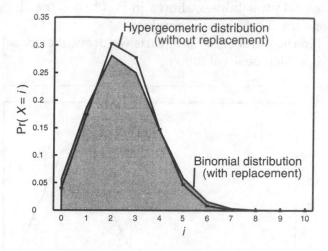

Figure 5.5: Choosing 10 cards from a deck of 52 cards with 13 hearts; determining the probability of drawing i hearts

replacement, which means that once an item has been selected, it cannot be selected again. Now consider a different sampling procedure: *sampling with replacement*. This time, draw a card, write it down, replace the card in the deck, re-shuffle the deck, and then draw the next card. Again, choose 10 cards, but this time each card might be drawn more than once. In fact, it is possible (although unlikely) that the same card might be drawn all 10 times.

Which sampling method is better? We can expect sampling without replacement to provide a slightly more accurate sample, since then we are guaranteed of choosing 10 different cards. However, it might not make much difference which method we use. Figure 5.5 compares the probabilities for drawing a 10-card sample with and without replacement. As before, the hypergeometric distribution gives the probabilities when the sample is chosen without replacement. We need to specify three numbers: $N = 52$ cards in the population; $M = 13$ hearts in the population; and $n = 10$ cards in the sample.

When the sample is chosen with replacement, then another distribution called the *binomial distribution* applies. The shaded area in Figure 5.5 shows the binomial distribution probabilities. We need to specify two numbers: $n = 10$ cards in the sample, and $p = .25$, the proportion of hearts in the population. You might be wondering: Why do we care about the difference between these two distributions? The answer is that the binomial distribution is easier to calculate, so it would be a big help if we could use it instead of the hypergeometric distribution.

Note that the hypergeometric distribution probabilities are concentrated closer to the peak, which means that sampling without replacement provides a slightly better sampling procedure. That is, there is a greater probability that the proportion of hearts in the sample will match the proportion of hearts in the population. However, there does not seem to be a big difference between the two methods, and in fact the difference becomes smaller and smaller as the population becomes larger and larger. Suppose we took four decks and shuffled them all together, giving us $N = 208$ cards with $M = 52$ hearts. The binomial distribution and hypergeometric distribution probabilities are compared in Figure 5.6; you can see there is very little difference between the two. In general, if the population is much larger than the sample, it doesn't matter whether the sample is chosen with or without replacement. The reason is that there is little probability of selecting the same item twice if the population is large. This is precisely the situation we often deal with in statistics; for example, if you are sampling 1,000 people from a population of 250 million, then clearly the population is much larger than the sample. Therefore, we will treat our samples as if they are chosen with replacement, even if they really have been chosen without replacement. (If you are unsure about whether this approximation is valid in a particular case, see **finite population correction factor** in Part II.)

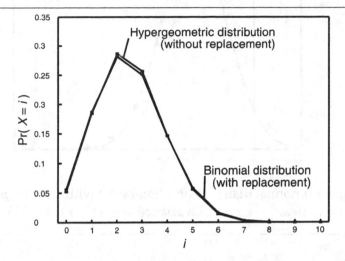

Figure 5.6: Choosing 10 cards from a deck of 208 cards with 52 hearts; determining the probability of drawing i hearts

THE NORMAL DISTRIBUTION APPROXIMATION

When you look at the graph of the binomial distribution in Figure 5.6, you might be tempted to wonder if we could make things even easier on ourselves by using the *normal distribution* instead. Figure 5.7 compares the binomial distribution for our card selection example ($n = 10, p = .25$) with the normal distribution. Recall that we must specify the mean and the variance in order to use the normal distribution. We need to use a normal distribution that has the same mean and variance as the corresponding binomial distribution.

Binomial distribution	Mean	Variance	Standard deviation
general	np	$np(1-p)$	$\sqrt{np(1-p)}$
with $n = 10, p = .25$	2.5	1.875	1.369

Note two important differences between the binomial and normal distributions. The binomial distribution is a discrete distribution that only takes on whole number values, whereas the normal distribution can take on any value within a continuous range (see **discrete random variable** and **continuous random variable** in Part II). Also note that the normal curve extends to the left into negative values of i. The number of hearts in the sample cannot be negative, so this is a weakness of the normal distribution approximation.

Despite these two problems, however, it seems clear from the graph that the normal distribution does a fairly reasonable job of representing the binomial distribution probabilities.

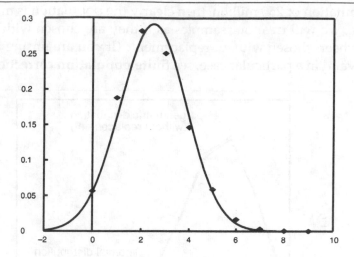

Figure 5.7: Comparing binomial distribution (diamonds) with $n = 10, p = .25$ with normal distribution curve with mean $= np = .25$, variance $= np(1 - p) = 1.875$

The normal distribution does an even better job of representing the binomial distribution if the sample size becomes larger. Figure 5.8 compares the binomial distribution probabilities (the diamonds) with the normal distribution curve with $n = 50, p = .25$. (See **Central Limit Theorem** in Part II for more information on when you can use the normal distribution as an approximation for other distributions.)

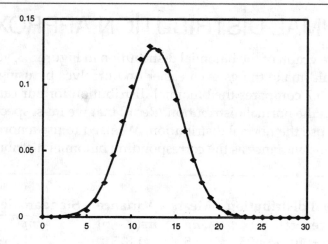

Figure 5.8: Comparing binomial distribution (diamonds) with $n = 50$, $p = .25$ with normal distribution curve with mean $= np = 12.5$, variance $= np(1 - p) = 9.375$

THE POPULATION AND SAMPLE PROPORTIONS

So far we have talked about cards, but the same ideas work if we are choosing a random sample from a population of people (or any other kind of object). Roughly 25% of the people in the U.S. are under age 17. Let $p = .25$, which is called the *population proportion*. Choose a random sample of size n from this population, and let X be the number of people in the sample under 17. Let $\hat{p}$ (read "p-hat") equal X/n, which is called the *sample proportion*. If we are lucky, then $\hat{p}$ will be close to p, meaning that our sample will be representative of the population. Under the right conditions (discussed above), X will have a normal distribution. Because of the properties of the normal distribution, $\hat{p}$ will also have a normal distribution:

	Mean	Variance	Standard deviation
X	np	$np(1-p)$	$\sqrt{np(1-p)}$
$\hat{p} = \dfrac{X}{n}$	p	$p\dfrac{(1-p)}{n}$	$\sqrt{\dfrac{p(1-p)}{n}}$

Note that the variance of $\hat{p}$ decreases as n increases, which means that the probabilities will be more closely concentrated around the true population proportion p as the sample size is increased. Figure 5.9 shows that your sample would not do a good job of representing the population proportion if $n = 10$. There is a fairly good chance that our sample proportion could be as large as .5 or as small as zero; in either case our sample would be wildly unrepresentative.

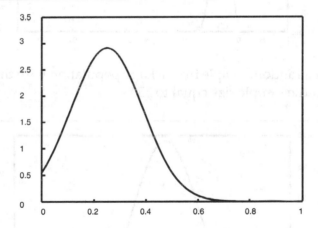

Figure 5.9: Choosing a random sample from a large population with the population proportion equal to 0.25, and the sample size equal to 10

The situation is much better with a sample of size 50 (see Figure 5.10) or size 250 (see Figure 5.11). Note how the probability density function reaches a higher, narrower peak about the true population proportion as the sample size increases. (The numbers on the vertical scale of these diagrams do not have special significance; recall that with a continuous random variable the important quantity is the area under the curve between two points. In all of these cases the total area under the curve is 1.)

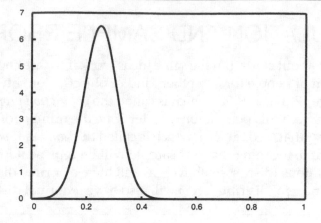

Figure 5.10: Choosing a random sample from a large population with the population proportion equal to 0.25, and the sample size equal to 50

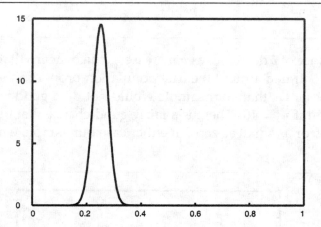

Figure 5.11: Choosing a random sample from a large population with the population proportion equal to 0.25, and the sample size equal to 250

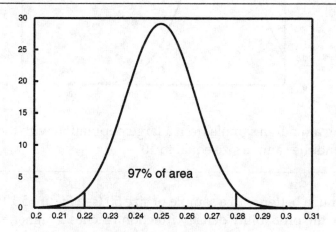

Figure 5.12: Choosing a random sample from a large population with the population proportion equal to 0.25, and the sample size equal to 1,000

If we really want to rest assured that our sample will reasonably represent the population, we should increase the sample size to about 1,000. Figure 5.12 shows that there is a 97% chance that our sample proportion will be between .22 and .28 with a sample size of 1,000. This is not good enough if we need an extremely precise sample (for example, if we need the sample proportion to be within .01 of the population proportion). However, for many practical purposes we can be satisifed if the chance of the sample proportion being within .03 of the population proportion exceeds 95%, so a sample of 1,000 will suffice.

CONFIDENCE INTERVAL FOR THE UNKNOWN POPULATION PROPORTION

There is one big problem with everything we have done so far in this unit: we have assumed we *know* the true population proportion. This is reasonable when we're talking about a small deck of cards, but unfortunately in the real world the true population proportion typically remains forever unknown. The whole point of taking a sample in that case is to use the sample proportion as an estimator of the population proportion. Our investigation so far gives us some confidence that a sample of 1,000 will give us a reasonable estimate of the unknown population proportion, but we need to put this in more precise terms.

We could ask: What is the probability that the sample proportion $\hat{p}$ will exactly equal the population proportion p? The chance of this happening is almost zero. The key word here is "exactly"; however, this turns out to be unnecessary because we only need to have the sample proportion come close. What we will do is create a little interval whose width is $2c$. This interval will be centered on $\hat{p}$, so its lower limit is $\hat{p} - c$ and its upper limit is $\hat{p} + c$. This interval is a random interval, because it depends upon the value of $\hat{p}$ that comes from our random sample. Imagine you are throwing a dart at a clothesline. You are guaranteed that the dart will hit the clothesline, but you don't know where on the line it will hit. The dart has little wings on each side (each of length c). The place where the center of the dart hits the clothesline is $\hat{p}$; you are trying to hit the true population proportion p. If we are lucky, the true value of p will be within the interval.

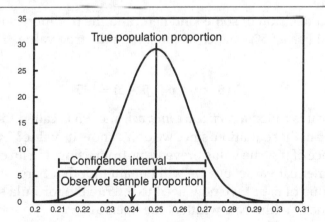

Figure 5.13: Choosing a random sample from a large population with the population proportion equal to 0.25, and the sample size equal to 1,000

Figure 5.13 illustrates a confidence interval where $\hat{p}$ is .24, and c is .03. Therefore, the interval is from .21 to .27, and the true population proportion of .25 is in the interval. However, in reality the normal curve and the true population proportion will be invisible to us. Figure 5.14 shows another situation with $\hat{p}$ equal to .24, only this time, unbeknownst to us, the true population proportion is .28. In this case we are unlucky because our confidence interval does not contain the true value. You could object that this is unlikely; you can see from the position of the normal curve that it would be unlikely that our sample proportion would be .24 if the population proportion is .28. We can't rule it out, however.

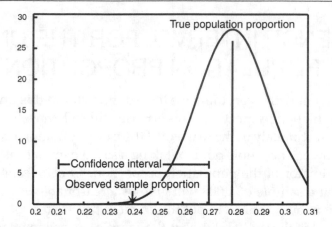

Figure 5.14: Choosing a random sample from a large population with the population proportion equal to 0.28, and the sample size equal to 1,000

"How do we decide the value of c?" you're probably wondering. If we make c very large, then we will have a wide interval that is almost certain to contain the true value of p. For example, we could make c equal to infinity, in which case we can be certain that the interval from $\hat{p} - \infty$ to $\hat{p} + \infty$ contains the true value. However, an interval that wide is totally useless.

On the other hand, if we make the value of c too small, then we have a narrower interval, which carries with it the disadvantage that there is a higher probability the interval might not contain the true value.

We have to make a decision of some kind here, and the tradition is to choose a value of c such that the interval has a 95% chance of containing the true value. This can be written as an equation:

$$\Pr(\hat{p} - c < p < \hat{p} + c) = .95$$

This kind of interval is called a *confidence interval*, and .95 is called the *confidence level*.

"How can we put p in that equation, since we don't know its value?" you might object. It is true that our ignorance of p's true value prevents us from using it in an equation that requires calculation of its numerical value, but we will not attempt that here. All the above formula does is define the value of c, but we now need to rewrite the formula so that it allows us to find a numerical value of c (the half-width of the interval).

$$c = 1.96\sqrt{\frac{\hat{p}(1 - \hat{p})}{n}}$$

(See the appendix at the end of the unit for the derivation of this formula.)

> **Formula for the 95% Confidence Interval for the Proportion**
> p = population proportion (unknown) that is, the fraction of items in the population that are in the category of interest
> n = sample size
> $\hat{p}$ = sample proportion (fraction of items in the sample in the category of interest)
> The 95% confidence interval is between these two limits:
> $$\hat{p} - 1.96\sqrt{\frac{\hat{p}(1-\hat{p})}{n}} \quad \text{to} \quad \hat{p} + 1.96\sqrt{\frac{\hat{p}(1-\hat{p})}{n}}$$

EXAMPLE 1

In a random sample of 900 Romans, 522 people supported Julius Caesar's assassination. Determine the 95% confidence interval for the true population proportion.

Calculate $\hat{p} = 522/900 = .58$. Then calculate c:

$$
\begin{aligned}
c &= 1.96\sqrt{\frac{\hat{p}(1-\hat{p})}{n}} \\
&= 1.96\sqrt{\frac{.58\times(1-.58)}{900}} \\
&= 1.96\sqrt{.00027067} \\
&= 1.96 \times .0165 \\
&= .032
\end{aligned}
$$

The confidence interval is .58±.03, or .55 to .61. We can expect that if we took 100 such polls, then 95 times our resulting confidence interval would contain the true population proportion.

For other examples of confidence intervals, see Unit 6.

WARNINGS

By now you should have developed a reasonable faith in a poll's ability to predict the population proportion. In the case of elections, we have a way of testing whether the poll is accurate, because the entire voting population is queried on election day. In many other cases polls are used in situations where we never know the true population characteristics.

However, don't become too complacent. There are several reasons why a poll designed to predict an election can be wrong. Here is a list of some of these, arranged roughly in decreasing order according to the amount you should worry about them:

1. The sample may not have been selected randomly. This is a crucial point—none of these statistical formulas work on a sample that is not randomly selected. How do we tell if the sample is random? Think of the analogy of shuffling a deck of cards: in a well-shuffled deck, you are equally likely to draw any card; not only that, if you are drawing 10 cards, then all possible 10-card hands are equally likely to appear. You can think of ways that this would not be true; for example, if some meticulous person has put the

deck in order and then you deal it without shuffling, it was not random. If you just collect the cards from the previous game and deal them without shuffling, it is not random because the act of playing the game has rearranged the cards in subtle (or sometimes obvious) ways.

If you are sampling people, then you cannot simply take a "convenience" sample—that is, ask the people that are most accessible to you. These people may differ systematically from people whom you can't reach.

You should attach no weight to psuedo-polls, such as when a television station has people call in to express an opinion. In that case there is not even a slight effort made to generate a random sample, so the people who call in are not representative of anything but themselves. (See **sampling** in Part II for more information on ways of selecting samples.)

2. Some people may refuse to cooperate with your poll. (Fortunately, this is not usually a problem if you are sampling inanimate objects.) If the people who refuse to answer differ systematically in their opinions from those who do answer, your sample will be biased. It is very difficult to measure this kind of bias for the very reason that you don't know the opinions of the people who don't tell you their opinions.

3. Your sample ideally should be chosen only from among those who actually will vote, but it is difficult to screen nonvoters out of your sample. You can ask if they intend to vote, but people may be reluctant to admit it if they're not intending to vote.

4. People can change their minds, so even if your poll is perfectly accurate the day it is taken, it will fail to predict the election if the voters change their minds before the election.

The Julius Caesar example mentioned previously illustrates the importance of timing. Suppose this poll were taken after the assassins' speech but before Mark Antony's speech. Do you think we could rely on these results to predict popular opinion?

5. The way the question is worded can affect the response. For example, your response to the question "Should the government continue to waste vast amounts of money on program X?" is likely to be different from your response to the question "Should program X continue to meet the pressing needs of people in our community?" The order in which the questions are asked can also matter. If you preface questions about a candidate by giving information about that candidate's stands on issues, you can make it more likely that the respondent will have a favorable (or unfavorable) impression of the candidate. A reputable poll-taking firm will try to state its questions in as neutral a fashion as possible. You should be very suspicious of any poll that comes directly from a candidate or advocacy group.

6. People may not respond honestly to the questions. If there is some kind of stigma attached to a particular viewpoint, then the respondent may give the socially acceptable answer rather than reveal true feelings. A good election poll will have the respondent fill out a secret ballot and put it in a box, so the interviewer does not know how the person votes. Unfortunately, you can't do this with a telephone poll.

7. The sample may be too small. Some polls try to get by with samples containing only a few hundred voters. If it is an honest poll, it will include a disclaimer of this general

form: "This poll was conducted by a random telephone sample of 900 registered voters. There is a 95% chance that the results given in this sample will be within about 3 percentage points of the true population proportion." A good poll will typically have about a thousand respondents, and will typically have a plus-or-minus error of $c = .03$ (3 percentage points). In a few cases the sample will be much larger than this. For example, the monthly unemployment figure comes from a massive survey called the Current Population Survey with more than one hundred thousand respondents. When the government reports the unemployment rate, they don't want to allow a 3% error; for example, they don't want to say the unemployment rate is 6% plus or minus 3%. Therefore, they need a very large sample to have a narrow confidence interval. Private poll takers do not have the resources to have samples this large, but you should be very suspicious of any poll with less than 500 respondents.

8. You may have had bad luck with the sample. There is a 5% chance that your 95% confidence interval will not contain the population proportion, even if you have done everything else right. Another way to put it: if you make a career out of taking polls, then you would expect that about 5 out of every 100 polls you do will be "wrong"— that is, the true value will not be within the confidence interval. The problem is that you won't know which 5.

APPENDIX TO UNIT 5: DERIVATION OF CONFIDENCE INTERVAL FORMULA

Start with the equation defining the confidence interval:

$$\Pr(\hat{p} - c < p < \hat{p} + c) = .95$$

Subtract $\hat{p}$ from all three parts of the inequality:

$$\Pr(-c < p - \hat{p} < c) = .95$$

Multiply all three parts by -1 (which requires reversing the directions of the inequalities):

$$\Pr(c > \hat{p} - p > -c) = .95$$

Change the order to return it to the customary form, where the inequalities point to the left:

$$\Pr(-c < \hat{p} - p < c) = .95$$

Recall that $\hat{p}$ has a normal distribution with mean p. The center part of the inequality has a normal random variable minus its mean; in similar situations before we also found it helpful to divide by the standard deviation:

$$\Pr\left(\frac{-c}{\sigma_{\hat{p}}} < \frac{\hat{p} - p}{\sigma_{\hat{p}}} < \frac{c}{\sigma_{\hat{p}}}\right) = .95$$

Here $\sigma_{\hat{p}}$ is the standard deviation of $\hat{p}$, which we found earlier to be $\sqrt{\frac{p(1-p)}{n}}$. Now make these definitions:

$$a = \frac{c}{\sigma_{\hat{p}}}$$

$$Z = \frac{\hat{p} - p}{\sigma_{\hat{p}}}$$

Then we can rewrite the equation as:

$$\Pr(-a < Z < a) = .95$$

The variable Z has the standard normal distribution, since it is formed by taking a normal random variable, subtracting its mean, and then dividing by its standard deviation. Therefore we can solve for a: $a = 1.96$ (see the standard normal table, page 205). Our mission was to solve for c, the halfwidth of the confidence interval, which we can do by rewriting the equation for a:

$$c = a\sigma_{\hat{p}} = 1.96\sqrt{\frac{p(1-p)}{n}}$$

Unfortunately, we do not know the value of p to insert into the formula, but it turns out that substituting $\hat{p}$ in its place gives us a reasonable approximation.

$$c = a\sigma_{\hat{p}} = 1.96\sqrt{\frac{\hat{p}(1-\hat{p})}{n}}$$

EXERCISES

1. Suppose you are investigating what fraction of the purchase orders of a certain firm are missing the part number. You have checked a sample of 100 purchase orders and found that 5% of them were missing the part number. Calculate a 95% confidence interval for the fraction of purchase orders in the entire population that are missing the part number.

2. Suppose you conduct a survey of 400 people and find that 40% of them have heard of your product. Calculate a 95% confidence interval for the fraction of people in the entire population that have heard of your product.

3. Suppose the people in a large city are about evenly divided between quiche lovers and quiche haters. You will select a random sample as part of a quiche preference survey. How many people must you include in the sample if you want to be able to calculate a 95% confidence interval for the population proportion that is 6 percentage points wide? How many people must be in the sample if you want the width of the confidence interval to be 2 percentage points?

4. What value of $\hat{p}$ will give you the widest confidence interval (other things being equal)?

For Exercises 5 to 12, you are given the value of n. Calculate the value of c (which is half of the width of the confidence interval) for confidence levels of .90, .95, and .99. Assume that $\hat{p} = .5$. For confidence level .90, use 1.65 instead of 1.96; for confidence level .99, use 2.58 instead of 1.96.

5.   10

6.   50

7.   100

8.   500

9.   1,000

10.   5,000

11. 10,000

12. 50,000

ANSWERS

1. $.05 \pm 1.96\sqrt{\frac{.05(1-.05)}{100}} = .05 \pm .043$ (.007 to .093)

2. $.40 \pm 1.96\sqrt{\frac{.4(1-.4)}{400}} = .40 \pm .048$ (.352 to .448)

3. $2 \times 1.96\sqrt{\frac{(.5)(1-.5)}{n}} = .06$

$4 \times 1.96^2 \left(\frac{.25}{n}\right) = .06^2$

$4268.44 = \frac{n}{.25}$

$n = 1,067$

For the confidence interval to have width .02, let $n = 9604$.

4. $\hat{p} = .5$

	Confidence level:	.90	.95	.99
	n	c	c	c
5.	10	0.261	0.310	0.408
6.	50	0.117	0.139	0.182
7.	100	0.083	0.098	0.129
8.	500	0.037	0.044	0.058
9.	1,000	0.026	0.031	0.041
10.	5,000	0.012	0.014	0.018
11.	10,000	0.008	0.010	0.013
12.	50,000	0.004	0.004	0.006

UNIT 6

Hypothesis Testing: Unknown Population Parameters

EXAMPLES OF HYPOTHESIS TESTING PROBLEMS

Two problems now confront us.

Dice Problem: A suspicious-looking stranger has possibly altered the numbers on the faces of a die. He won't allow us to see the numbers on all faces, but we will be able to toss the die and observe the number that appears. What chance do we have of determining whether the die really has been altered?

Fish Problem: A large number of fish live in a lake. We would like to know if the lengths of these fish average 8 in., as claimed by an eager vacation time-share salesperson. If we could observe all the fish in the entire population, then we could calculate the average of those numbers (call it μ). However, we have no way of measuring all of the fish, so we will have to content ourselves with observing a randomly selected sample. Our natural inclination is to use the average length of fish in the sample as an estimator of the average length of fish in the population; the question is—will this be a reliable estimator?

These are just two examples of the dilemmas that statisticians face in dealing with the real world. Things were easier in probability theory, where we deal with a known process and try to predict the outcomes.

Probability

In statistical inference, the problem is reversed. We observe the outcomes (for example, the numbers that appear when the die is tossed, or the length of fish in the sample) and try to predict the nature of the process (whether it is a fair or altered die, and the length of fish in the entire population).

Statistical Inference

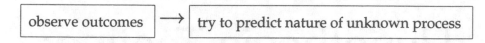

DICE PROBLEM

Our malicious stranger has a die with six numbers (call them $a_1, a_2, a_3, a_4, a_5, a_6$), which may or may not be the standard numbers that are supposed to be on a die (1, 2, 3, 4, 5, 6). We will never be allowed to inspect all faces of the die, but we will see the result of tossing the die a certain number of times. We need to use the information we see on those tosses to decide whether or not the die is fair.

If we knew the numbers on the die, we could calculate the true mean μ:

$$\mu = \frac{a_1 + a_2 + a_3 + a_4 + a_5 + a_6}{6}$$

We could also find the true population variance:

$$\sigma^2 = \frac{(a_1 - \mu)^2 + (a_2 - \mu)^2 + (a_3 - \mu)^2 + (a_4 - \mu)^2 + (a_5 - \mu)^2 + (a_6 - \mu)^2}{6}$$

The quantities μ and σ are said to be *parameters* of the unknown distribution. They will remain unknown, but we can calculate the average and standard deviation of the sample we can observe. The quantities calculated from the sample are called *statistics*. A sample statistic that is being used as an indicator of the value of an unknown parameter is called an *estimator*. An estimator is a special type of random variable. It is random because its value depends on the result of a random process (in this case, which items appear in the sample). Sometimes a hat is placed over a quantity to indicate it is an estimator; for example, $\hat{\mu}$ would be one notation for an estimator for μ.

Suppose we are only allowed a sample of size 1. If we let X_1 be the number that appears when we toss the die once, we can calculate the expected value $E(X_1)$ by multiplying each possible value by its probability:

$$E(X_1) = \frac{1}{6}a_1 + \frac{1}{6}a_2 + \frac{1}{6}a_3 + \frac{1}{6}a_4 + \frac{1}{6}a_5 + \frac{1}{6}a_6$$

This works out to be the same as μ.

The fact that the expected value of X_1 equals μ means that X_1 is said to be an *unbiased estimator* of μ. In general:

An estimator $\hat{A}$ is an *unbiased estimator* of a parameter A if $E(\hat{A}) = A$.
In words: the expected value of an unbiased estimator is equal to the
true value of the parameter it is trying to estimate.

The property of being unbiased is a very desirable feature in an estimator. It means that if
you repeat the estimation process many times, on average you will be right. However, just
determining that an estimator is unbiased does not necessarily make it a good estimator. We
also need to consider how much variability it has. If we toss the die once, then there is a lot of
uncertainty in our estimator. It would be much better to toss the die many times and calculate
the average of the numbers that appear. Such an estimator will also be unbiased, and it will
have a smaller uncertainty (making it a better estimator).

Now let's look for an unbiased estimator for the population variance σ^2. If we knew the
true value of μ, we could use this estimator:

$$\hat{\sigma^2} = (X_1 - \mu)^2$$

This is an unbiased estimator of σ^2. However, we don't know μ, so we will have to use the
sample average $\bar{x}$ instead. This immediately creates a problem; in a sample of size 1, $\bar{x}$ is
the same as X_1, which gives us an estimator of the variance that is zero. We don't have any
confidence in this result; obviously there is no variance in the sample if the sample only has
one element in it. However, there is a problem that applies even when we take larger samples.
If we have a sample of size 2 (X_1 and X_2) then this is an unbiased estimator for σ^2:

$$\hat{\sigma^2} = \frac{(X_1 - \mu)^2 + (X_2 - \mu)^2}{2}$$

Unfortunately, this estimator again suffers from the fatal defect that it requires knowledge of
μ. Therefore, we will try this estimator, which uses the sample average $\bar{x} = \frac{(X_1 + X_2)}{2}$ in place
of μ:

$$s_1^2 = \frac{(X_1 - \bar{x})^2 + (X_2 - \bar{x})^2}{2}$$

This estimator suffers from the disadvantage that it tends to underestimate the true σ^2. In
general, if we define s_1^2 using this formula:

$$s_1^2 = \frac{\sum_{i=1}^{n}(X_i - \bar{x})^2}{n}$$

it turns out that:

$$E(s_1^2) = \frac{(n-1)\sigma^2}{n}$$

Since $E(s_1^2)$ does not equal σ^2, it means that s_1^2 is not an unbiased estimator of σ^2. In order to
create an unbiased estimator, we will define a new estimator s^2:

$$s^2 = \frac{\sum_{i=1}^{n}(X_i - \bar{x})^2}{n-1}$$

This is the formula that will usually be used for the variance of a sample. Note that the formulas for s_1^2 and s^2 are the same except that $n - 1$, rather than n, appears in the denominator of s^2. If the sample is large, then there will not be very much difference between these two versions. The square root of s^2 is known as the *sample standard deviation*.

To illustrate how we would go about testing for the value of the unknown mean, we will now consider the fish situation.

TESTING FOR THE UNKNOWN MEAN FISH LENGTH

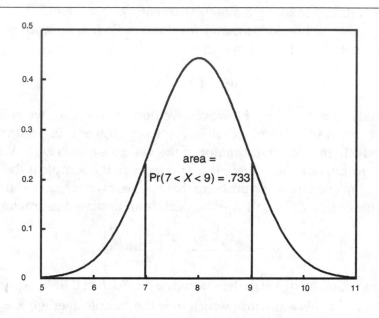

Figure 6.1

An eager vacation time-share salesperson promises us that the average length of trout in the lake is 8 in. The question is: Do we trust him?

If we are allowed to catch one fish before we decide, our procedure will be as follows: Measure the fish; if its length is close to 8 in., accept the claim; if it is far from 8 inches, reject the claim and accuse the salesperson of deception.

However, there is a definite risk to this strategy. Suppose we decide to accept the claim (called a *hypothesis*) if the length of the fish is between 7 and 9 in. Further, suppose we happen to know that the lengths have a normal distribution with standard deviation equal to $\sigma = 0.9$ in. Assume for now that the hypothesis is true, so the true mean does equal 8. Then we can calculate the probability that the length of one randomly selected fish (call it X) will be between 7 and 9 (see Figure 6.1):

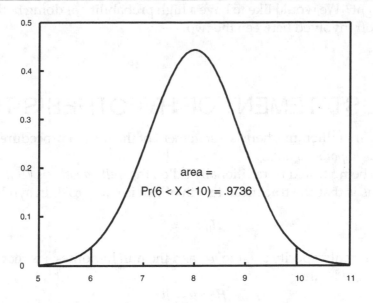

Figure 6.2

$$Pr(7 < X < 9) = Pr\left(\frac{7-8}{0.9} < \frac{X-8}{0.9} < \frac{9-8}{0.9}\right)$$
$$= Pr(-1.11 < Z < 1.11)$$
$$= Pr(Z < 1.11) - Pr(Z < -1.11)$$
$$= .8665 - .1335$$
$$= .7330$$

Therefore, we can be reasonably confident that if the true mean fish length really is 8, the size of the fish we catch will be between 7 and 9. However, the disturbing part is that there is still a $1 - .733 = .267$ probability that the length of the fish will be outside the range from 7 to 9 (even if the true mean is 8). If that happens, we would falsely accuse an innocent salesperson.

We could reduce the probability of this happening by widening our acceptance range. For example, we might decide that we will accept the hypothesis that the true mean is 8 if the length of our randomly selected fish is between 6 and 10. Then the probability of being within our acceptance range if the hypothesis is true increases (see Figure 6.2):

$$Pr(6 < X < 10) = Pr\left(\frac{6-8}{0.9} < \frac{X-8}{0.9} < \frac{10-8}{0.9}\right)$$
$$= Pr(-2.22 < Z < 2.22)$$
$$= Pr(Z < 2.22) - Pr(Z < -2.22)$$
$$= .9868 - .0132$$
$$= .9736$$

Now there is only a chance of $1 - .9736 = .0264$ that we will reject the hypothesis if it is really true. We might lose a little sleep over this probability, but we will reluctantly conclude that we are going to have to live with some nonzero probability of accusing an innocent salesperson. As long as the probability is low, we will figure that most likely justice will be served.

At the same time, however, we now raise the opposite problem: if we widen our acceptance region, then we are increasing the likelihood that a crooked salesperson will be trusted. We face a tough choice here—is it more important to make sure the innocent are protected or the guilty are caught? We would like to have a high probability of doing both, but there is an element of trade-off involved between the two.

FORMAL STATEMENT OF HYPOTHESIS TESTING

While you ponder that dilemma, here is a statement of the formal procedure in the branch of statistics known as *hypothesis testing*.

The hypothesis being tested is traditionally called the *null hypothesis* (H_0). In our example, the null hypothesis is that the true mean length of fish is equal to 8. In symbols:

$$H_0 : \mu = 8$$

More generally, we could specify a value $\mu*$, and the null hypothesis is that the true value μ is equal to the specified value:

$$H_0 : \mu = \mu*$$

(The reason for calling it the null hypothesis is not clear in this example, but it does become clear in some examples such as testing medicines. In that case the null hypothesis states that the medicine has no effect.)

Our mission is to accept or reject the null hypothesis. If we reject the null hypothesis, then we accept the statement "the null hypothesis is false," which is, logically enough, known as the *alternative hypothesis*. However, accepting the hypothesis does not mean that we have proved it is true; it just means that the evidence found so far seems to be consistent with it.

There are two different types of errors that we can make in hypothesis testing:

- rejecting the null hypothesis when it is really true (this is called a type I error)

- accepting the null hypothesis when it is really false (this is called a type II error)

For example, if we had accused the salesperson when the true mean really was 8, then we would have committed a type I error. On the other hand, if we had accepted the hypothesis when the mean was some other number (say 7.2) then we would have committed a type II error.

The following table summarizes the errors:

	H_0 is correct	H_0 is wrong
accept H_0	right	type II error
reject H_0	type I error	right

The probability of committing a type I error is called the *level of significance* of a test procedure. For example, our first test procedure (accepting the hypothesis if the fish length was between 7 and 9) had a significance level of .267. The second procedure (where the acceptance region was from 6 to 10) had a significance level of .0264. Note that a *lower* significance level is better if you are trying to avoid a type I error. Another way to put it is that if the significance level is low, then a decision to reject the null hypothesis is more meaningful, since you know there is a low probability that you will falsely reject it.

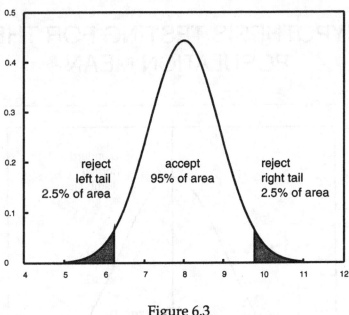

Figure 6.3

The typical procedure in hypothesis testing works like this:

1. Decide on the significance level to use. Traditionally, 5% is the most common level. Another popular choice is 1%.

2. Decide on a test statistic to use. The test statistic chosen must meet this condition: *if* the null hypothesis is true, it will come from a known distribution. In our example, the test statistic was the length of the fish that we caught; we know that if the true mean was 8, then the length of that fish would act as a random number chosen from a normal distribution with mean 8 and standard deviation 0.9.

3. Decide on an acceptance zone and a rejection zone for the test statistic. For our example (with $\sigma = 0.9$ and a 5% significance level), the acceptance zone would be $8 \pm 1.96\sigma = 8 \pm 1.764$, which is from 6.236 to 9.764 (see Figure 6.3). (The rejection zone is also called the *critical region*.)

4. Conduct your observations and calculate the test statistic value. If the test statistic falls within the acceptance zone, then you will accept the hypothesis (at least for now, until further evidence comes along). Otherwise you will reject the hypothesis. If you conducted your test with a low significance level, then you can be confident that the hypothesis is wrong if you are led to reject it. We can make it precise by asking: How low can you make the significance level go and still reject the hypothesis? This number is called the *p* value of the test.

HYPOTHESIS TESTING FOR THE POPULATION MEAN

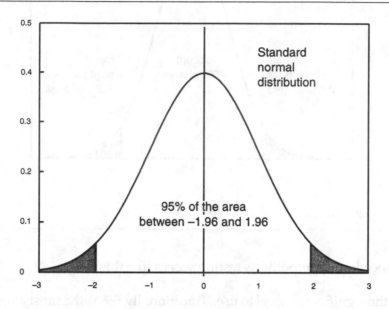

Figure 6.4: 2.5% of area in each tail

There is one major flaw in our fish story up to now—we would never want to generalize about the entire population of fish if our sample consisted of only a single fish. Obviously we would be much better off if we caught a sample of n fish.

If $\bar{x}$ is the average length of fish in a random sample of size n, chosen from a normally distributed population of fish with mean μ and standard deviation σ, then we know that $\bar{x}$ has a normal distribution with mean μ and standard deviation $\sigma/\sqrt{n}$ (see page 55.) (Recall how the standard deviation of $\bar{x}$ becomes smaller as n becomes larger; statisticians would give up in despair if this fact were not so.)

If the null hypothesis is true, then the true value μ equals the hypothesized value $\mu*$, and we therefore know that the distribution of $\bar{x}$ is normal with mean $\mu*$, standard deviation $\sigma/\sqrt{n}$. We could use $\bar{x}$ directly as the test statistic, but it is helpful to convert it into the form of a standard normal random variable Z.

$$Z = \frac{\bar{x} - \mu*}{\sigma/\sqrt{n}} = \frac{\sqrt{n}(\bar{x} - \mu*)}{\sigma}$$

See page 61 where we did similar transformations.

Set up the acceptance zone as shown in Figure 6.4. There is a 5% chance that a standard normal random variable will be outside the range -1.96 to 1.96, so that will be the rejection region.

USING THE *T* DISTRIBUTION

There is a big flaw, however, with the analysis so far: we assumed that σ, the standard deviation of the population of fish, was known. Most likely this will not be the case; after all, you could not know σ unless you could investigate the entire population, which would mean you would also know the true value of μ. We realize our only choice is to reach for the nearest available substitute—the standard deviation estimated from the sample.

We need to modify the test statistic Z:

$$Z = \frac{\bar{x} - \mu*}{\sigma/\sqrt{n}} = \frac{\sqrt{n}(\bar{x} - \mu*)}{\sigma}$$

We will create a test statistic T that is almost the same, except that the sample standard deviation s is substituted in place of the unknown population standard deviation σ:

$$T = \frac{\bar{x} - \mu*}{s/\sqrt{n}} = \frac{\sqrt{n}(\bar{x} - \mu*)}{s}$$

We would expect that the distribution of T will be close to a standard normal distribution, but it won't be exactly the same. The distribution of this statistic is known as the *t distribution*. In order to use the t distribution, we need to specify a quantity called the *degrees of freedom*. For this situation, there are $n - 1$ degrees of freedom.

We need to look in the t distribution table (page 209) to determine the acceptance zone and rejection zone. If our sample is size 15, then we look in the table for $15 - 1 = 14$ degrees of freedom, and we find:

$$\Pr(-2.145 < T < 2.145) = .95$$

See Figure 6.5.

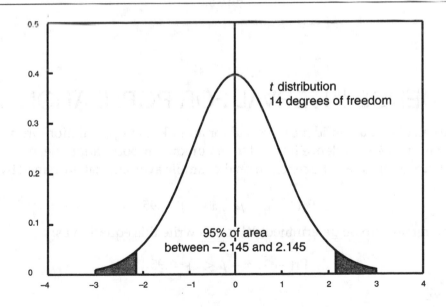

Figure 6.5

For example, suppose we randomly catch 15 fish with these results:

7.5, 8.2, 8.1, 8.4, 7.1, 7.3, 7.1, 7.8, 8.0, 7.3, 7.3, 7.9, 7.8, 8.1, 7.6

Our computer will calculate $\bar{x} = 7.70$ and $s = 0.41918$. Calculate the test statistic:

$$\frac{\sqrt{15}(7.7 - 8)}{0.41918} = -2.77$$

This falls within the rejection region, so we do not accept the hypothesis that the true mean is 8 (see Figure 6.6).

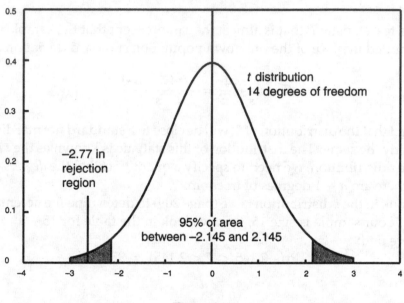

Figure 6.6

CONFIDENCE INTERVAL FOR POPULATION MEAN

In Unit 5 we calculated a confidence interval for an unknown population proportion; now we can calculate a 95% confidence interval for an unknown population mean.

The confidence interval will be centered on the sample average $\bar{x}$; it is defined by this equation:

$$\Pr(\bar{x} - c < \mu < \bar{x} + c) = .95$$

The value of c remains to be determined. We can rewrite this equation as:

$$\Pr(-c < \bar{x} - \mu < c) = .95$$

Multiply by $\frac{\sqrt{n}}{s}$:

$$\Pr\left(\frac{-c\sqrt{n}}{s} < \frac{\sqrt{n}(\bar{x} - \mu)}{s} < \frac{c\sqrt{n}}{s}\right) = .95$$

Let $a = \frac{c\sqrt{n}}{s}$ and $T = \frac{\sqrt{n}(\bar{x}-\mu)}{s}$; then

$$\Pr(-a < T < a) = .95$$

We can look up the value of a in the t distribution table with $n - 1$; as we have seen, for 14 degrees of freedom $a = 2.145$. Then we can rewrite the formula so we can solve for c:

$$c = \frac{as}{\sqrt{n}}$$

Note that c becomes smaller as n becomes larger. This makes the confidence interval narrower and more precise. This is good, but it might be too expensive to obtain a larger sample. The confidence interval becomes wider if s becomes bigger, meaning there is greater spread within the sample. This is bad, but there is not much we can do about it.

For our fish example, ($\bar{x} = 7.70$ and $s = 0.41918.$). We find that $c = \frac{2.145 \times 0.41918}{\sqrt{15}} = 0.23$. Therefore, the 95% confidence interval for the fish length is $7.70 \pm .23$, which is from 7.47 to 7.93.

Following are two more examples with different samples:

Example 1	Example 2
7.5	7.5
7.2	8.2
6.6	8.1
6.8	8.4
7.1	8.1
7.3	7.3
7.1	7.1
7.8	7.8
7.0	8.0
7.3	8.3
7.3	7.3
7.9	7.9
7.8	7.8
6.8	8.1
7.6	7.9
Total: 109.100	Total: 117.800
Average: 7.27333	Average: 7.85333
s: 0.39182	s: 0.39073
0.950 confidence interval:	0.950 confidence interval:
7.056 to 7.490	7.637 to 8.070

For Example 1, we would use this test statistic to test the hypothesis that μ equals 8:

$$\frac{\sqrt{15}(7.27 - 8)}{0.392} = -7.2$$

Since $-7.2 < -2.145$, this is well within the rejection region, so for this sample we would reject the hypothesis that the true mean is 8.

For Example 2, we would use this test statistic:

$$\frac{\sqrt{15}(7.85 - 8)}{0.391} = -1.49$$

Since -1.49 is between -2.145 and 2.145, this is within the acceptance region, so for this sample we would accept the hypothesis that the true mean is 8.

GENERAL STATEMENT: HYPOTHESIS TESTING AND CONFIDENCE INTERVALS FOR MEANS

In general: Let $X_1, X_2, \ldots X_n$ be a random sample selected from a population with mean μ and standard deviation σ. Assume that this population has a normal distribution; or, alternately, assume that n is large enough that $\bar{x}$ can be assumed to have a normal distribution by the Central Limit Theorem. Assume either that the population is much larger than the sample, or that the sample is selected with replacement. Calculate the sample average:

$$\bar{x} = \frac{X_1 + X_2 + X_3 + \ldots + X_n}{n}$$

Calculate the sample standard deviation:

$$s = \sqrt{\frac{(X_1 - \bar{x})^2 + (X_2 - \bar{x})^2 + (X_3 - \bar{x})^2 + \ldots + (X_n - \bar{x})^2}{n - 1}}$$

Define the random variable T:

$$T = \frac{\sqrt{n}(\bar{x} - \mu)}{s}$$

Then T will have a t distribution with $n - 1$ degrees of freedom.

To test the hypothesis that μ equals a specified value $\mu*$, calculate this test statistic:

$$T = \frac{\sqrt{n}(\bar{x} - \mu*)}{s}$$

This statistic will have a t distribution with $n - 1$ degrees of freedom if the null hypothesis is true. Reject the hypothesis at the 5% significance level if the calculated test statistic is outside the range $-a$ to a, where a is found from the t distribution table (page 209) such that

$$\Pr(-a < T < a) = 0.95$$

To determine a 95% confidence interval for the mean, look up a as shown above, and then use these formulas for the end points of the interval:

$$\bar{x} \pm \frac{as}{\sqrt{n}} \text{ or } \bar{x} - \frac{as}{\sqrt{n}} \text{ to } \bar{x} + \frac{as}{\sqrt{n}}$$

Note from page 209 that the t distribution comes closer and closer to the standard normal distribution as the number of degrees of freedom increases. If the sample is larger than 60, then you can typically use the standard normal random table instead; the t distribution is reserved for situations where you are dealing with small samples.

For more information, see entries in Part II on **analysis of variance; chi-square test; goodness-of-fit test; p value; power of test;** and **two-tailed hypothesis test.**

EXERCISES

For Exercises 1–9, you are given a set of observations from a random sample taken from a population with a normal distribution. First, calculate a 95% confidence interval for the mean. Then, test the hypothesis that μ equals the given value of $\mu*$, using the 5% significance level.

1. 10, 6, 15, 25, 17; $\mu* = 15$

2. 110, 190, 156, 135, 170; $\mu* = 140$

3. 2, 6, 4, 7, 5; $\mu* = 6$

4. 23, 21, 29, 25, 21, 26, 24, 27; $\mu* = 32$

5. 21, 36, 15, 17, 27, 46, 36, 41; $\mu* = 60$

6. 60, 85, 74, 96, 56, 70, 87, 78; $\mu* - 50$

7. 90, 80, 70, 100, 85, 90, 100, 80, 60, 90, 80; $\mu* = 76$

8. 56, 63, 75, 54, 58, 43, 59, 43, 47, 60, 56; $\mu* = 50$

9. 87, 79, 67, 46, 87, 54, 67, 43, 56, 33, 57; $\mu* = 97$

For Exercises 10–15, you are given values for $\bar{x}$, s, n, and $\mu*$. Calculate a 95% confidence interval for μ, assuming the values of $\bar{x}$ and s come from a random sample. Then test the hypothesis that $\mu = \mu*$.

	$\bar{x}$	s	n	$\mu*$
10.	120	14	50	118
11.	640	168	80	680
12.	1285	142	100	1100
13.	36	15	200	35
14.	94	36	300	98
15.	76	8	500	74

ANSWERS

For Exercises 1–9, calculate the sample average $\bar{x}$ and the sample standard deviation s. The degrees of freedom (DF) is $n-1$; the value of a comes from the t distribution table. The value of c comes from the formula $c = as/\sqrt{n}$. The 95% confidence interval is from $\bar{x} - c$ to $\bar{x} + c$. To test the hypothesis that $\mu = \mu*$, calculate the test statistic (TS) from the formula $TS = \sqrt{n}(\bar{x} - \mu*)/s$. Accept the null hypothesis if the absolute value of the test statistic is less than the value of a from the t distribution table.

	$\bar{x}$	s	DF	a	c	$\bar{x} - c$	$\bar{x} + c$	TS	
1.	14.60	7.232	4	2.776	8.978	5.62	20.22	−0.124	accept
2.	152.20	30.971	4	2.776	38.449	113.75	265.95	0.881	accept
3.	4.80	1.924	4	2.776	2.388	2.41	7.21	−1.395	accept
4.	24.50	2.828	7	2.365	2.365	22.14	46.64	−7.500	reject
5.	29.88	11.544	7	2.365	9.653	20.22	50.10	−7.381	reject
6.	75.75	13.636	7	2.365	11.401	64.35	140.10	5.341	reject
7.	84.09	12.004	10	2.228	8.064	76.03	160.12	2.236	reject (barely)
8.	55.82	9.282	10	2.228	6.236	49.58	105.40	2.079	accept
9.	61.45	17.801	10	2.228	11.958	49.50	110.95	−6.623	reject

	$\bar{x}$	s	c	$\bar{x} - c$	$\bar{x} + c$	TS	
10.	120	14	3.9	116.1	236.1	1.010	accept
11.	640	168	36.8	603.2	1243.2	−2.130	reject
12.	1285	142	27.8	1257.2	2542.2	13.028	reject
13.	36	15	2.1	33.9	69.9	0.943	accept
14.	94	36	4.1	89.9	183.9	−1.925	accept
15.	76	8	0.7	75.3	151.3	5.590	reject

UNIT 7

The Relationship Question: Regression Analysis

Now it is time to discuss the relationship question. Specifically, how can we tell if two quantities are related to each other? In other words, do they move together, so that one quantity is large when the other is large (and vice versa)? Another way to word it: Does knowledge of one of the quantities help you predict the value of the other?

Here are some sample questions:

- Is there a relationship between the two numbers that appear when two dice are tossed? (Your guess: probably not.)

- Is there a relationship between the number on the top of a die and the number on the bottom of a die? (Guess yes.)

- Is there a relationship between the area of a state and its population? (Maybe.)

- Is there a relationship betwen a state's population and its electoral votes? (Definitely yes.)

There are three types of situations we will investigate:

1. where we understand the process generating the quantities, as in the dice example, where we can use probability theory. Unfortunately, this will not often be the case with the problems we face in statistics, but we can still learn from these situations.

2. where we have data on the complete population. For example, we can collect data from all 50 states, so we can settle once and for all the question about the nature of the relationship of quantities involving states.

3. where we have data from a sample selected from the population. The usual warning applies: we can't even begin to apply statistical analysis if the sample is not randomly selected. We will investigate the relationship between the quantities in the sample, and then see whether it is likely that the relationship generalizes to the entire population.

SCATTERPLOTS

To analyze relationships, we first need to collect some data (or determine the outcomes in the case of a probability problem).

EXAMPLE 1

Let X be the number that appears on the first die, and let Y be the number that appears on the second die. There are $6 \times 6 = 36$ total possible outcomes for these two random variables; we can make a table called a *joint probability table* showing the probability of each of these outcomes:

Y \ X:	1	2	3	4	5	6	TOTAL
1	1/36	1/36	1/36	1/36	1/36	1/36	1/6
2	1/36	1/36	1/36	1/36	1/36	1/36	1/6
3	1/36	1/36	1/36	1/36	1/36	1/36	1/6
4	1/36	1/36	1/36	1/36	1/36	1/36	1/6
5	1/36	1/36	1/36	1/36	1/36	1/36	1/6
6	1/36	1/36	1/36	1/36	1/36	1/36	1/6
TOTAL	1/6	1/6	1/6	1/6	1/6	1/6	1

(The row totals give the probability for the random variable Y by itself; these are called the *marginal probabilities* of Y. Likewise, the column totals give the marginal probabilities of X.)

We can illustrate this relationship with a scatterplot or scatter graph, where each mark represents one of the possible outcomes (see Figure 7.1).

The answer to the relationship question in this case is clearly no. Knowledge of X provides you with no help in predicting the value of Y.

EXAMPLE 2

Now, let X be the number on top of the die, and let Y be the number on the bottom of the die. (It is traditional to use X and Y for the two variables you are comparing, so you will have to get used to X and Y meaning different quantities in different examples.) It happens that dice are constructed so that the numbers on opposite faces add up to 7, so we can construct the joint probability table as follows:

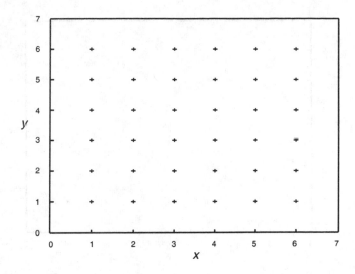

Figure 7.1: X = number on first die; Y = number on second die

Y \ X:	1	2	3	4	5	6	TOTAL
1	0	0	0	0	0	1/6	1/6
2	0	0	0	0	1/6	0	1/6
3	0	0	0	1/6	0	0	1/6
4	0	0	1/6	0	0	0	1/6
5	0	1/6	0	0	0	0	1/6
6	1/6	0	0	0	0	1/6	1/6
TOTAL	1/6	1/6	1/6	1/6	1/6	1/6	1

Note that the marginal probabilities here are the same as in the previous example with two dice, but the joint probability table is completely different. These outcomes can be illustrated on a scatter graph (see Figure 7.2), which shows that there is a strong relationship between these two variables. If you know the value of X, then you can easily determine the value of Y.

Now toss three dice. Let X be the sum of the first two dice; let Y be the sum of the second and third dice. You would expect there to be some relationship between these two cases, since they both depend on the value that appears on the middle die. However, the first and third dice are independent of each other, so knowing X will not allow you to make perfect predictions of Y. That is often the way of the world—there is some relationship, but not a perfect relationship.

We can construct a table showing the possible outcomes. Note that $(1, 1, 1)$ is the only outcome where $X = 2$ and $Y = 2$; but there are six outcomes where $X = 7$ and $Y = 7$: $(1, 6, 1)$, $(2, 5, 2)$, $(3, 4, 3)$, $(4, 3, 4)$, $(5, 2, 5)$, $(6, 1, 6)$.

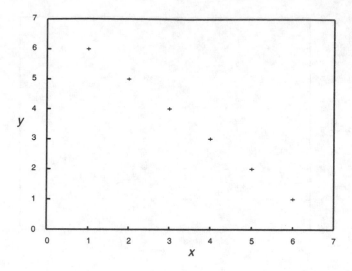

Figure 7.2: X = number on top of die; Y = number on bottom of die

Y \ X:	2	3	4	5	6	7	8	9	10	11	12	TOTAL
2	1	1	1	1	1	1	0	0	0	0	0	6
3	1	2	2	2	2	2	1	0	0	0	0	12
4	1	2	3	3	3	3	2	1	0	0	0	18
5	1	2	3	4	4	4	3	2	1	0	0	24
6	1	2	3	4	5	5	4	3	2	1	0	30
7	1	2	3	4	5	6	5	4	3	2	1	36
8	0	1	2	3	4	5	5	4	3	2	1	30
9	0	0	1	2	3	4	4	4	3	2	1	24
10	0	0	0	1	2	3	3	3	3	2	1	18
11	0	0	0	0	1	2	2	2	2	2	1	12
12	0	0	0	0	0	1	1	1	1	1	1	6
TOTAL	6	12	18	24	30	36	30	24	18	12	6	216

From this we can construct a table of probabilities (in each case, this is the number of outcomes divided by 216).

Figure 7.3 shows the scatter diagram. (In cases where more than one mark would appear on the same spot, the subsequent marks have been shifted slightly.)

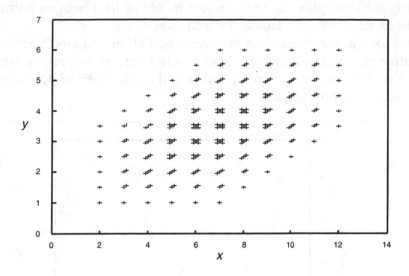

Figure 7.3: X = number on first two dice; Y = number on last two dice

EXAMPLE 3

Do states larger in area have more people than smaller states? In this case we cannot describe the probabilities associated with the process that has determined the sizes of the states, but we do know all of the numbers in the population, so we can create a scatter diagram (see Figure 7.4).

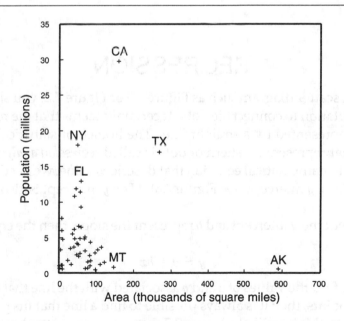

Figure 7.4

Alaska clearly is an exception to our suggested relationship. The other points suggest there may be a slight relation between area and population.

Next we will look at the connection between population and electoral votes. Figure 7.5 shows the scatter diagram, which indicates there is a very strong relationship. We would expect to find this, since electoral votes are allocated on the basis of population (even if the formula is not strictly proportional).

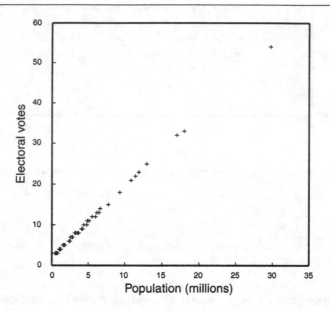

Figure 7.5

REGRESSION

When you look at a scatter diagram such as Figure 7.5 or Figure 7.2, you should be filled with the irresistible temptation to connect the dots. It certainly seems that the relationship in those two cases can be represented by a straight line. The branch of statistical analysis involved with fitting lines that represent a pattern of dots is called *regression analysis*.

In order to write a mathematical equation that describes a line, we need to know two numbers: the slope, and the y intercept (see Figure 7.6). (The y intercept is also called the constant term).

If we let a represent the y intercept and b represent the slope, then the equation can be written:

$$y = a + bx$$

Our mission is to find the values of a and b associated with the line that best fits the data. If there are only two points, then it is always possible to find a line that fits perfectly; if there are more than two points, this is unlikely. Figure 7.7 shows that no matter how hard you try, there is some deviation between the points and the line you are using to try to represent them. We want to choose the line that minimizes these deviations (called *errors*.) The normal procedure is to square the errors, add them up, and minimize the result.

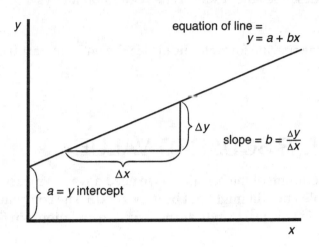

equation of line =
$$y = a + bx$$

Δy

slope $= b = \dfrac{\Delta y}{\Delta x}$

Δx

$a = y$ intercept

Figure 7.6

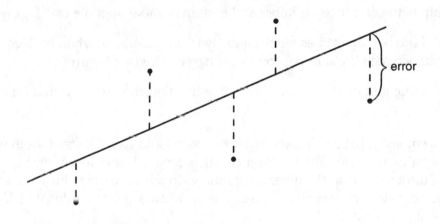

error

Figure 7.7

This sounds like a lot of work, but fortunately we can turn it over to the computer. Here is the general procedure for regression analysis:

1. Assemble your observations for the two quantities. Each observation of x must be matched with one observation of y; for example, they could both come from the same state, or the same person, or the same year.

2. Input the data into a computer program, such as a spreadsheet program or statistics program.

3. Have the computer create a scatter graph of the data. This step is important because it allows you to visualize whether or not there seems to be a relationship.

4. Call the command that performs a simple regression analysis on the data. (It is called simple because there is only one independent variable; later we will use multiple regression analysis.)

5. The computer will report back values for the slope and the intercept, along with some other information that describes the result of the regression analysis (to be discussed shortly).

(For more information on the specific formulas used, see the entry in Part II on **simple regression**.)

THE *R* SQUARED VALUE

We need more than just a description of the best line—we need a way to measure whether it is very good. (The best possible line still might not be very good.) The computer will report a value known as the r squared value (r^2) to indicate how well the relationship fits. These are the properties of the r^2 value:

1. r^2 is always between 0 and 1.

2. $r^2 = 1$ if all of the observations fit along a straight line, as when we looked for the relationship between the top number and bottom number on a die (see Figure 7.2).

3. $r^2 = 0$ if the two quantities are completely independent, as when we looked for a relationship between the numbers on two different dice (see Figure 7.1).

4. The r^2 value gives the percent of variation in y that can be accounted for by variations in x.

Suppose a contest is to be held between Rosencrantz and Guildenstern. Both will be trying to guess the value of a variable y. Rosencrantz has no information except for the average value of y. Guildenstern, on the other hand, knows in advance the value of x, and he knows the regression equation connecting y and x. The question is: Will Guildenstern do better than Rosencrantz at guessing y?

Rosencrantz's best strategy is simply to guess that y will be equal to its average $\bar{y}$. For any given observation of y (call it y_i), there will be a certain amount of error between the actual value and Rosencrantz's prediction:

$$\text{error}_i = y_i - \bar{y}$$

To get the total error of Rosencrantz's plan, we will add these up (squaring the error first, so all of the errors become converted into positive numbers). Call this quantity SE_{avg}, for squared error about the average:

$$SE_{avg} = \sum_{i=1}^{n}(y_i - \bar{y})^2$$

This assumes that there are n observations.

Guildenstern might have an advantage over Rosencrantz because his guess for y will come from the regression equation:

$$\hat{y}_i = a + bx_i$$

As before, the hat placed over the y indicates that it is an estimator. Let e_i represent the error that Guildenstern will make from his prediction:

$$e_i = y_i - \hat{y}_i = y_i - (a + bx_i)$$

We can calculate the total squared error of Guildenstern's predictions, calling it SE_{line}, for squared error about the regression line:

$$SE_{line} = \sum_{i=1}^{n}(y_i - \hat{y}_i)^2 = \sum_{i=1}^{n}[y_i - (a + bx_i)]^2$$

(See Figure 7.8.)

Suppose that Guildenstern's squared error using the regression is just as large as Rosencrantz's, who did not use the regression. In that case the regression is totally worthless. Knowledge of x provides no help in predicting the value of y. On the other hand, suppose that SE_{line} is zero. In that case, Guildenstern is able to make perfect predictions using the regression analysis, and Rosencrantz doesn't stand a chance in the competition.

Therefore, we define the r^2 value as follows:

$$r^2 = 1 - \frac{SE_{line}}{SE_{avg}}$$

So, $r^2 = 1$ if SE_{line} is 0, and $r^2 = 0$ if $SE_{line} = SE_{avg}$.

One other possibility to consider is to suppose that the observations of x and y fit along a perfectly horizontal line. You would be tempted to say there is a relationship, because you can find a line that fits all the points. However, because the slope of this line is zero, it is clear that changing the value of x doesn't affect the value of y. It turns out in this case that you don't need x to predict the value of y, because y never changes. The r^2 value is undefined because it would require division by zero.

EXAMPLES OF SIMPLE REGRESSION CALCULATIONS

Here are some examples.

- Numbers on two different dice (see Figure 7.1): There is no way to draw a line representing this relationship. The regression result will report a slope of 0, a y intercept or constant of 3.5, and an r^2 value of 0.

- Numbers on the top and bottom of a die (see Figure 7.9): Note the perfect fit.

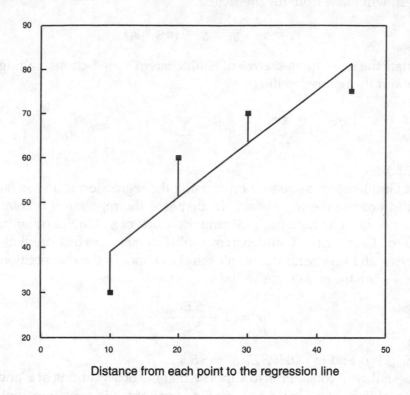

Distance from each point to the regression line

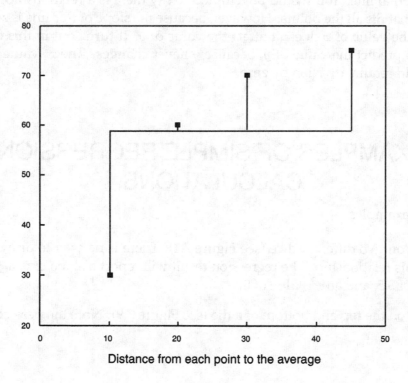

Distance from each point to the average

Figure 7.8

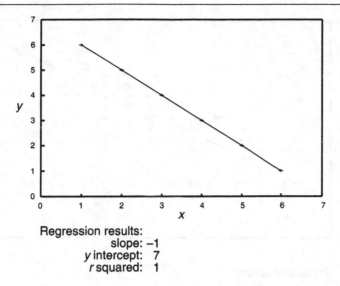

Regression results:
slope: −1
y intercept: 7
r squared: 1

Figure 7.9: X = number on top of die; Y = number on bottom of die

- X = sum of the first two of the three dice; Y = last two of the three dice (see Figure 7.10): Note that only 25% of the variation in Y is explained by variation in X.

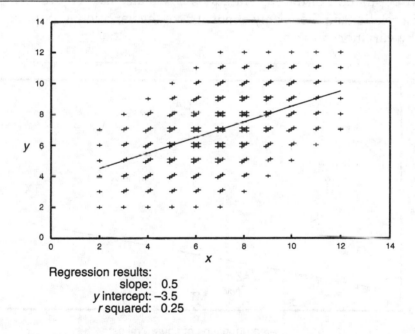

Regression results:
slope: 0.5
y intercept: −3.5
r squared: 0.25

Figure 7.10: X = sum of numbers on first two dice; Y = sum of numbers on last two dice

- X = sum of the first two of the three dice; Y = the sum of all three numbers: We would expect that in this case our regression should explain more than it did in the previous example; this is confirmed in Figure 7.11.

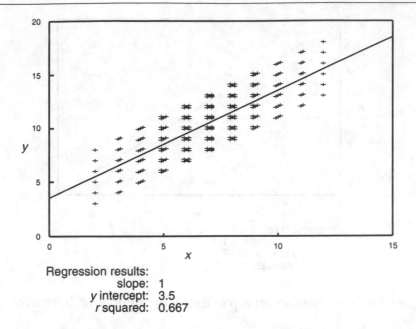

Regression results:
       slope: 1
  y intercept: 3.5
    r squared: 0.667

Figure 7.11: $X = $ sum of numbers on first two dice; $Y = $ sum of numbers on three dice

- Area and population of the 48 contiguous U.S. states (see Figure 7.12): There is not much of a relationship here; only about 10% of the variation in population can be explained by variations in area.

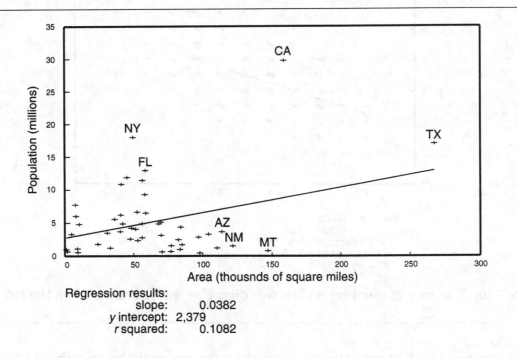

Regression results:
       slope: 0.0382
  y intercept: 2,379
   r squared: 0.1082

Figure 7.12

- Population and electoral votes (see Figure 7.13): As expected, there is a strong relationship. The r^2 is very close to 1. It is not exactly 1 because electoral votes are not exactly proportional to population.

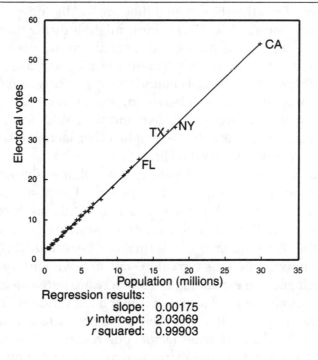

Figure 7.13

STATISTICAL ANALYSIS OF REGRESSION

So far we have considered examples where we know the process generating the variables (dice), or where we know data for the complete population (the states). More commonly, we only have data from a sample, and we will use our analysis of the relationship visible in that sample to predict a relationship that would be visible in the population. Sometimes our sample will be limited because of time. We have no way of observing data from the future, so we are limited to using observations from the present and past if we want to discover a relationship between the variables. If the future relationship will stay the same as it was in the past (except for the same level of random variation), then we can use our past observations to predict the future relationship. Other times we may be looking at data from a sample of people (or other type of object) to try to predict properties of the entire population.

Our assumption in regression analysis is that there exists a relationship between two variables x and y that can be described by this equation:

$$y = a + bx + e$$

Here a and b are unknown parameters, x is called the *independent variable*, and y is the *dependent variable*. The assumption is that variations in x are responsible for causing the variation in y. However, you must be careful with this assumption, since the mere fact you have established a relationship between x and y does not mean that x causes the changes in y. It might be that y causes the changes in x, or it could be that there is a third unidentified factor that causes changes in both x and y. (For example, any two variables that both grow with time will appear to have a relationship, even if they are totally independent.) Or you might even have bad luck with your sample. Your sample might indicate there is a relationship between x and y when in fact there is no such relationship in the population. The chance of that happening is small if the sample is large enough, but if you make a career of performing regression analysis that kind of bad luck is bound to happen occasionally.

The quantity e is a random variable called the *error term*. If x were the only variable that affected y, then no error term would be needed, and we could find a perfect fit to our regression line. However, there almost always will be other factors affecting y that we don't observe. If we don't have any information about these, we have to assume that their effect can be described as random chance. The assumption is that e is a random variable with a normal distribution, 0 mean, and unknown variance σ^2. The assumption of 0 mean is not restrictive. (If by some chance e had nonzero mean, then this mean could be added to the constant a, which would redefine e as a new random variable that would have zero mean.) The assumption that the distribution of e is normal can be tested. It is unfortunate that the true value of σ^2 is unknown, but, as is typical in statistics, we will try to estimate it. If this regression is to be much good in explaining the relationship between x and y, then σ^2 needs to be relatively small. Another way of saying it is that if the random variable e contributes a large part of the variance of y, then it means there are other factors influencing y in addition to x, and you somehow should track those factors down and include them in your analysis. ("What if we find there is more than one quantity that affects y?" you might be wondering. Look ahead to the section on multiple regression.)

After you have collected the observations for your sample and fed them into the computer, the computer will return the results of the regression calculation—the slope and constant of the regression line. However, these values are not necessarily the same as the true values a and b. We would know those true values only if we could observe the entire population. Instead, we use the regression coefficients from the sample as estimators for unknown parameters. That means we can perform hypothesis tests on these estimators (see **t statistic in regression** and **F statistic in regression** in Part II for examples).

CORRELATION COEFFICIENT

Another quantity used to describe the strength of a relation is the *correlation coefficient*, which is related to the r^2 value:

- The r^2 value is the square of the correlation coefficient (which is known as r).

- The sign of the correlation coefficient is the same as the sign of the slope of the regression equation. If the slope is positive, the correlation is positive; if the slope is negative, the correlation is negative.

- The value of the correlation coefficient ranges from –1 to 1, with 1 indicating a perfect linear relationship with a positive slope, –1 indicating a perfect linear relationship with a negative slope, and 0 indicating no linear relationship.

(See also **covariance** in Part II.)

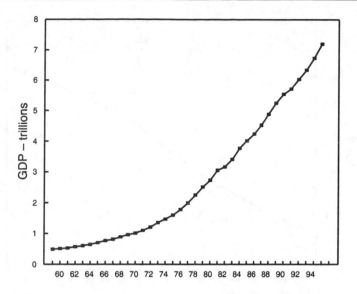

Figure 7.14: Regression r squared: 0.9244

CURVED RELATIONSHIPS: LOGARITHMIC TRANSFORMATIONS

You probably feel that our focus on linear relationships is too restrictive. We might well expect to find perfectly healthy relationships that don't fit the linear mold. Figure 7.14 shows the growth of gross domestic product (GDP) for the years 1959 to 1995. This is an example of a common type of regression analysis where time is used as the independent variable. If a quantity grows along a fairly steady trend, this is a good way to predict its future growth.

A simple regression of GDP and time gives an r^2 value of .9244, which indicates a good fit. However, the graph indicates that we would have a better fit if we could somehow represent the relation as a curve. We can do that with a transformation: take the logarithm of the GDP. (See the entry on **logarithm** in Part II if you need to review the concept.) Figure 7.15 shows the graph of the logarithm of the GDP over time; you can see that it is much closer to a straight line. The r^2 value is .9935, and the regression equation is:

$$\log(GDP_t) = 11.62168 + 0.0350t$$

The variable t represents time, defined so that the year 1959 is called time 1, 1960 is time 2, and so on. The quantity GDP_t represents the value of the GDP at time t.

To create an equation for the original value of the GDP (not its logarithm), we need to perform an exponentiation—take 10 raised to the power of the left and right hand sides of the equation:

$$
\begin{aligned}
10^{\log(GDP_t)} &= 10^{11.62168 + 0.0350t} \\
GDP_t &= 10^{11.62168} \times 10^{0.0350t} \\
GDP_t &= (4.185 \times 10^{11}) \times 1.084^t
\end{aligned}
$$

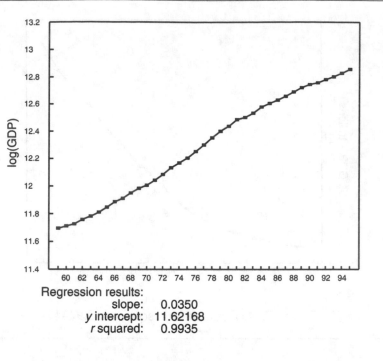

Regression results:
         slope: 0.0350
    y intercept: 11.62168
    r squared: 0.9935

Figure 7.15

From this equation we can find that:

$$\frac{GDP_{t+1}}{GDP_t} = \frac{(4.185 \times 10^{11}) \times 1.084^{t+1}}{(4.185 \times 10^{11}) \times 1.084^{t}} = 1.084$$

In words, this states that every year, the value of the GDP will be 1.084 times as great as it was the previous year, or that it grows at 8.4% per year.

This type of transformation works for any quantity that grows at roughly a constant percentage rate per year:

$$y_t = y_0(1+g)^t$$
$$\log y_t = \log(y_0) + t \log(1+g)$$

The expression $\log(y_0)$ represents the constant term in the regression analysis, using $\log y_t$ as the dependent variable; the quantity $\log(1+g)$ represents the slope; and g represents the growth factor (the percentage growth rate divided by 100).

If the scatter graph indicates that the relationship can be described by a U-shaped curve (as in Figure 7.16) or an upside-down U, then use multiple regression with both x and x^2 as independent variables.

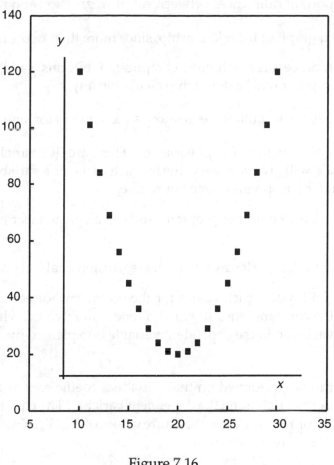

Figure 7.16

MULTIPLE REGRESSION

Simple regression analysis, as we have discussed so far, includes only one independent variable and one dependent variable. Sometimes, though, one independent variable simply isn't enough. For example, suppose that the price of a house (Y) in a particular city depends on four variables: square feet in house (X_1); distance to business center (X_2); distance to nearest school (X_3); and the interest rate (X_4). Assume that the dependent variable Y is related to the four independent variables according to an equation of this form:

$$Y = B_0 + B_1 X_1 + B_2 X_2 + B_3 X_3 + B_4 X_4 + e$$

The quantity B_0 is the constant term, analogous to the y intercept term in simple regression, and B_1, B_2, B_3, and B_4 are called the coefficients. The true values of all of the B's are unknown; we will estimate them based on our regression calculation. The letter e again represents an error term; assume that it has a normal distribution with mean 0 and unknown variance σ^2. All factors affecting house prices other than the four we have included are represented by e. The value of σ^2 should be relatively small; otherwise, there are important factors influencing house prices that we have not included, and our regression equation will not be able to do a very good job of predicting house prices.

There are two important differences between simple regression and multiple regression:

- We can't draw a graph of the relationship, since more than two variables are involved.

- The calculation process is much more complicated, but this is not a serious drawback because the computer will be doing the calculation anyway.

Much of the procedure for multiple regression is the same as for simple regression:

1. Assemble your observations for all variables. There are five variables in our example. Each observation will give you one value for each of the five variables. In our example, each observation corresponds to one house sale.

2. Input the data into a computer program, such as a spreadsheet program or statistics program.

3. Call the command that performs a multiple regression analysis on the data.

4. The computer will report back values for the coefficient for each of the independent variables and the constant term. It will also report an r^2 value, which again tells what percentage of variation in the dependent variable is explained by the regression equation.

Hypothesis tests can be performed on the individual coefficients to test the relationship between an independent variable and the dependent variable. The equation can also be used to predict future values of Y if you know the future values of X_1, X_2, X_3, and X_4. See **multiple regression** in Part II for more information.

EXERCISES

For Exercises 1–5, you are given observations for two variables. Use a computer regression routine to calculate the slope, intercept, and r^2 value.

1.

	X1	Y1
1:	4	52
2:	23	200
3:	2	32
4:	15	162
5:	21	222
6:	3	42
7:	20	250
8:	14	152

2.

	X2	Y2
1:	9	15
2:	10	11
3:	0	3
4:	13	11
5:	8	10
6:	7	2
7:	12	14
8:	5	13

3.

	X3	Y3
1:	10	184
2:	4	88
3:	2	56
4:	5	104
5:	1	40
6:	7	136
7:	3	72
8:	2	56
9:	4	88
10:	7	136

4.

	X4	Y4
1:	67	138
2:	77	83
3:	45	47
4:	53	81
5:	58	43
6:	70	109
7:	84	76
8:	44	46
9:	55	72
10:	99	124

5.

	X5	Y5
1:	41	267
2:	49	305
3:	9	119
4:	21	174
5:	14	159
6:	34	249
7:	7	192
8:	8	114
9:	23	183
10:	22	271

ANSWERS

1. Slope: 9.702
Intercept: 15.304
r squared: 0.939

2. Slope: 0.692
Intercept: 4.342
r squared: 0.348

3. Slope: 16.000
Intercept: 24.000
r squared: 1.000

4. Slope: 1.208
Intercept: 2.818
r squared: 0.423

5. Slope: 3.920
Intercept: 113.925
r squared: 0.732

UNIT 8

Fallacies and Traps

Statistics can be a powerful force for good, as we have seen, but it can also create mischief if it is not applied correctly. This unit will discuss several different issues where, if you are not careful, you could incorrectly apply a statistical concept. Reading this unit will also help you gain the ability to expose sinister charlatans who deliberately use statistics to mislead people.

EXAMPLE 1

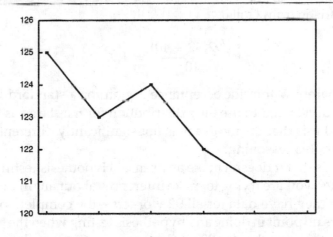

Figure 8.1

The XYZ corporation wishes to emphasize that they have cut the price of their widgets, as shown in Figure 8.1. The fall in price looks dramatic, until you look closely. Note the vertical scale of the graph—it starts just below the lowest data point, and the graph stops just above the highest data point. Any data series will appear to show a tremendous shift if you construct the graph in this fashion. The rule is that a graph of a change in a variable with time should always have a vertical scale that starts at zero, as in Figure 8.2. Otherwise it is inherently misleading. (Note: you may include a graph that does not start at zero as a supplement to illustrate a magnified view of the changes, as long as the reader can also see the nondistorted graph.)

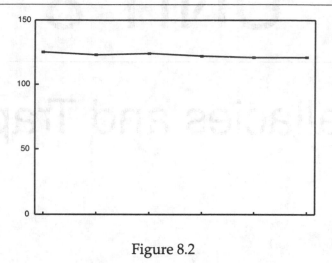

Figure 8.2

EXAMPLE 2

A small company wishes to investigate the ages of its workers. It has collected data from all 92 workers, and finds an average age of 42 with a standard deviation of 10. It then proceeds to see if a hypothesis testing procedure will show that the age is significantly different from 40, using this test statistic from Unit 6:

$$\frac{\sqrt{92}(42 - 40)}{10} = 1.92$$

It concludes this is barely within the acceptance region for a standard normal distribution (reasoning that they don't need to use the t distribution for n as large as 92.) Therefore, they will accept the hypothesis that the mean age is not significantly different from 40.

What's wrong with this reasoning?

The basic problem is that it doesn't make any sense. Hypothesis testing is a procedure applied to samples, when you are trying to make inferences about an unknown population parameter. In this case they have data for all 92 workers—the complete population—so they know $\mu = 42$. There is no point in doing any hypothesis testing when the parameter is known.

Perhaps they were thinking that the 92 workers constitute a sample from a larger population of workers, but statistical analysis still would not be applicable because the sample was not selected randomly.

EXAMPLE 3

A teacher with 7 students calculates the standard deviation of their test scores as follows:

$$\sqrt{\frac{(90-70)^2 + (80-70)^2 + (80-70)^2 + (70-70)^2 + (40-70)^2 + (60-70)^2 + (70-70)^2}{7-1}}$$

$$= 1.63299$$

What is wrong with this calculation? The teacher is using the formula for the standard deviation of a sample—except this data does not come from a sample. If you are interested in the performance of this class on the exam, then you know all the data for the population, so you will calculate the standard deviation of a population (with n, rather than $n-1$, in the denominator):

$$\sqrt{\frac{(90-70)^2 + (80-70)^2 + (80-70)^2 + (70-70)^2 + (40-70)^2 + (60-70)^2 + (70-70)^2}{7}}$$

$$= 1.51186$$

Perhaps the teacher was thinking of using this class as a sample of students chosen from the population of all students in the school, but again this is inappropriate because the sample is not selected randomly.

EXAMPLE 4

One day we were visiting New York City, and met a friend named Bill on the street. Out of all the people in New York City, it seems astounding that you could meet someone you know by chance. Suppose we somehow are able to estimate that the chance of meeting Bill is 1 in 100,000. Should we be truly astonished that such an unlikely event happened?

The question is: Why are we interested in calculating the probability of meeting Bill? If we really wanted to know this probability, we should have calculated it *before* we met Bill. Then it would have been truly astounding to meet Bill; in fact, we could say, "This is amazing, Bill! Just the other day, we were calculating the probability that we would meet you by chance on the street in New York City, and here you are!" However, this is not the way it happened. Instead, we did not become interested in the probability of meeting Bill until *after* we had already met him.

Now consider this question: What is the probability that we would meet someone—anyone—we know on the streets of New York? If we know 100 people in New York, then we might calculate that the probability of meeting someone we know is 100 times greater than the probability of meeting Bill specifically, so we would call the probability 1 in 1,000. Obviously it is much more likely that we will meet a generic person we know, rather than one specific person, such as Bill. Therefore, the amazing coincidence is not quite as amazing as it seems.

This is the key question you should ask whenever an amazing coincidence seems to have occurred: Did you wonder about the probability of this coincidence *before* or *after* it happened? If you truly identified the probability of a particular event as being very small, and then it happens, that is truly amazing. However, you will find life is full of amazing coincidences if you are allowed to consider the likelihood of occurrence after they have happened. For example, suppose you are driving down the street and note with astonishment that the license plate of the car ahead of you reads FLG 427. You realize that the chance of this happening is only 1 in $26^3 10^3 = 17,576,000$. However, that is only an astonishing coincidence if you had been wondering in advance whether you would be behind car FLG 427; otherwise, it is a totally pointless piece of trivia.

EXAMPLE 5

After taking a sample of four Democrats and four Republicans, you find the following values for broccoli consumption:

Democrats	Broccoli	Republicans	Broccoli
Bill	7	George	0
Jimmy	20	Ron	12
Lyndon	15	Jerry	15
John	10	Dick	11

You will determine if there is a correlation between political party and broccoli consumption, using the formula for correlation on page 152.

What is the problem?

This calculation doesn't make any sense, because you don't have matched observations for the variable. Try to make a scatter graph of the data, recalling that each point in a scatter graph can be labeled because it comes from one person, or one time period, or one something. If you plot the point ($x = 7, y = 0$), how do you label it? You can't label it, because the 7 comes from Bill, and the 0 comes from George.

What we could do is set up the data like this:

Person	Broccoli	Party
Bill	7	1
Jimmy	20	1
Lyndon	15	1
John	10	1
George	0	0
Ron	12	0
Jerry	15	0
Dick	11	0

Here we create a *dummy variable* to represent party, with 1 for Democrats and 0 for Republicans. Then we have 8 matched observations we could perform a regression on.

EXAMPLE 6

A researcher is investigating differences between men and women in a characteristic called x-ability. A random sample of $n = 1,600$ men and $n = 1,600$ women are selected; the sample average for men is $\overline{x_m} = 60.10$ and for women is $\overline{x_w} = 60.85$. The standard deviation of both samples is $\sigma = 10$. The researcher performs a hypothesis test for the difference between two means (described in the entry **difference of means** in Part II):

$$Z = \frac{\overline{x_m} - \overline{x_w}}{\sqrt{\frac{\sigma^2}{n} + \frac{\sigma^2}{n}}}$$

$$Z = \frac{60.10 - 60.85}{\sqrt{\frac{100}{1,600} + \frac{100}{1,600}}}$$

$$Z = \frac{-0.75}{\sqrt{0.125}} = -2.12$$

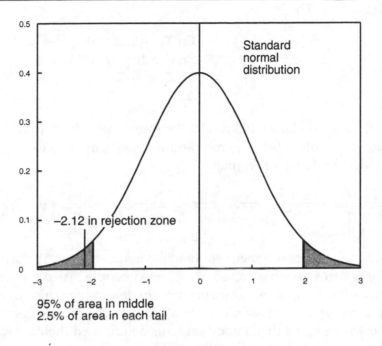

95% of area in middle
2.5% of area in each tail

Figure 8.3

The null hypothesis states there is no difference in mean x-ability between men and women. If this hypothesis is true, then Z will have a standard normal distribution. Since −2.12 is less than 1.96, we can reject the null hypothesis at the 5% significance level (see Figure 8.3). Therefore, the researcher can proclaim there is a statistically significant difference in x-ability between men and women.

The question remains: Should we be cautious when we discuss this result? Unfortunately, the terminology traditionally used in statistics does obscure what is important. The key word here is "significant." When we say that the result of this study is significant, it means merely that we can reject the hypothesis that the result happened by chance. In other words, we can use this evidence to believe that there truly is a difference in mean x-ability between men and women.

But just the fact that there is a difference does not mean it is a big difference. Here it is very important to remember that the word "significant" as used in statistics does *not* mean "important" as we would usually use the word. Someone not schooled in statistics would probably notice the obvious fact right up front—the difference between 60.10 and 60.85 is small, so even if there is truly a difference it is not an important difference. Furthermore, since the standard deviation for both men and women is equal to 10, we realize that there is much greater variability within each sex than there is between the sexes. If we tried to graph both normal curves, they would be practically indistinguishable.

One way to measure the difference precisely is to ask this question: What is the probability that a randomly chosen woman will have an x-ability score higher than a randomly chosen man? Assume that our sample mean and standard deviation figures can be used as good estimates of the population figures (a reasonable assumption, since we have large random samples). Let X_w be the x-ability score for our randomly chosen woman, and X_m be the x-ability score for our randomly chosen man. Assume X_w has a normal distribution with mean $\mu_w = 60.85$ and $\sigma = 10$; X_m has a normal distribution with mean $\mu_m = 60.10$ and $\sigma = 10$. Then $X_w - X_m$ has a normal distribution with mean $\mu_w - \mu_m = 0.75$ and standard deviation $\sqrt{\sigma^2 + \sigma^2} = \sqrt{200} = 14.1$. Then:

$$\begin{aligned} \Pr(X_w > X_m) &= \Pr(X_w - X_m > 0) \\ &= \Pr(Z > (0 - .75)/14.1) \\ &= \Pr(Z > -0.05) \\ &= .52 \end{aligned}$$

This result is close to .5, so you should not let the researcher's finding of a "statistically significant" difference in x-ability between men and women cause you to start stereotyping the x-ability of any individual man or woman.

EXAMPLE 7

A researcher trying to predict whether people will engage in type X behavior administers a 1,000 question survey to a group of randomly selected people. Then the researcher will run separate regressions for each of the 1,000 questions to determine which ones show a significant relationship with type X behavior. Is there a conceptual problem here?

To see the problem, imagine that the respondents determined their answer for each question by tossing coins—in other words, totally randomly. If you use the 5% significance level, then you would expect that 50 questions will show a significant connection to type X behavior, solely by chance. You must be very careful about attaching meaning to this result.

The problem here is similar to the example of meeting Bill on the streets of New York, discussed above. It would have been better to state the hypothesis about which questions matter in advance, then perform the regression analysis to test that hypothesis. If you engage in "data mining" or "data snooping"—that is, testing many regressions before stating any hypotheses—you have a good chance of mistaking coincidences for meaningful results.

Granted, it is hard to resist the temptation to state hypotheses after testing the data. Here is an important question to ask in that case: Can the results be replicated with a different sample? Any result in either the natural sciences or the social sciences should not be totally accepted until it can be shown to be replicated by different researchers with independent samples.

EXAMPLE 8

Someone hears the statement, "22% of poor people are fatherless children," but then repeats it as "22% of fatherless children are poor." The two statements sound similiar, but they say very different things. (In this case, the first one is right, according to the 1994 U.S. *Statistical Abstract*; the second one is incorrect.) To find the percent of poor people that are fatherless children, divide 8 million poor fatherless children by 36.9 million poor people, giving 22%.

To find the percent of fatherless children that are poor, divide 8 million poor fatherless children by 14.7 million fatherless children, giving 54%. Both of these statements are examples of conditional probabilities; it is important to make sure you state each one correctly. See **conditional probability** and **Bayes's rule** in Part II for more information.

EXAMPLE 9

A researcher calculates a 95% confidence interval for an unknown mean μ, and then states: "There is a 95% chance that the value of μ will be between 70 and 78." This is a subtle point, but this is not the correct statement of the meaning of a confidence interval. The mean μ is not a random variable, so it does not make sense to talk about the probability that it takes on certain values. Instead, it is the confidence interval itself that is random, so a correct statement would be: "There is a 95% chance that the random interval $\bar{x} - \frac{as_2}{\sqrt{n}}$ to $\bar{x} + \frac{as_2}{\sqrt{n}}$ will contain the unknown true value of μ." This is usually shortened to a saying involving the specific numbers, such as "the 95% confidence interval is from 70 to 78," which is fine as long as you remember the correct interpretation of the statement.

EXAMPLE 10

A researcher uses regression to find a strong relationship between x and y, and then claims this proves that x causes y. This is not necessarily true; it may be that y causes x, or it may be that a third variable causes changes in both x and y.

For example, a researcher observes data that the price of cars has increased over the last four decades, at the same time that the number of cars sold has increased, thereby concluding that higher prices for cars will cause more people to buy cars. This is very hard to believe; presumably people will buy more cars at lower prices, other things being equal. In reality, other things are not equal; for example, the price increase is caused largely by inflation, and the increase in demand is caused largely by population increases, among other factors. Any simple regression calculation can give misleading results if you have left out variables that should have been included. In that case you need to perform a multiple regression calculation, but even then there is a risk of getting misleading results because some variables have been left out.

EXAMPLE 11

Figure 8.4 shows a regression calculation performed on two variables x and y. There seems to be a good fit, and one is tempted to use this regression line to predict that if x becomes 20, then y will become slightly less than 300. The problem with this type of prediction is that we are going beyond the range of previously observed data. Such a prediction is called an *extrapolation*, and you must be very wary of it. For example, suppose that some more data is collected that follows the pattern shown in Figure 8.5. You can see how far off your prediction would be if it were based on the regression line that came from the limited sample.

A sad fact of life is that extrapolation predictions are what we often are very interested in. We would like to know how things would be different if we do something very different than we have ever done before. Since a relationship that is perfectly trustworthy within one range of observations may break down outside that range, we must be very careful of such predictions.

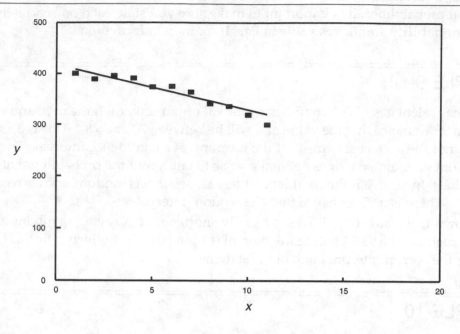

Figure 8.4

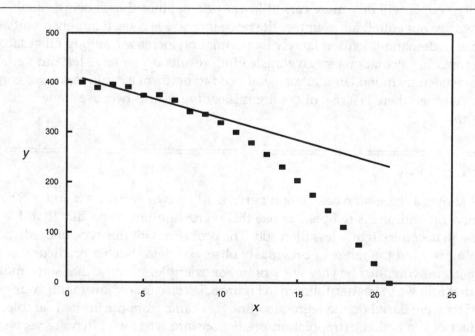

Figure 8.5

EXAMPLE 12

Finally, we must once again mention one of the most common traps you can fall into—applying statistical analysis to convenience samples. We have discussed the statistical magic that makes it possible to use results from a sample to predict results for a population, but that only works if the sample is randomly selected. That allows us to determine the probabilities that the sample will be like the population, so we can tell if the result could have happened by chance or not. You can't make that kind of generalization if the sample was not selected randomly.

A notable example of this was the *Literary Digest* poll of the 1936 presidential election. Even though the sample was very large (several million respondents), it was wildly inaccurate because it was a convenience sample. It predicted Alf Landon would win. The poll was selected from phone lists and magazine subscription lists—both of which systematically biased the sample. (At the time, higher income households were more likely to have phones or read the magazine.) The inaccuracy of this sample was exposed when the election was held and Franklin Roosevelt won. There are many other cases where statistical analysis is done on convenience samples but the population results are never checked, so the truth is not known. You can make a list of lots of ways not to select a sample: call your friends; ask people at a particular shopping mall; reply to a TV station call-in poll; respond to a magazine poll of its subscribers; or ask a class of college freshmen.

You have no way of knowing how the respondents might be different from the population, and you cannot rely on random selection to cancel out the bias if the sample is not selected randomly.

PART II

Alphabetical Reference
Section

Adjusted R Squared

A slightly different calculation of the r squared for a regression model that decreases as more independent variables are added unless they actually improve the model. If n is the number of data points, and k is the number of independent variables, then the adjusted r^2 is given by

$$1 - \frac{n-1}{n - (k + 1)}(1 - r^2)$$

The ordinary r^2 will increase by adding new independent variables, whether or not they belong in the regression, which creates the illusion that you have improved the model. In the extreme case, you could find a meaningless perfect fit (ordinary $r^2 = 1$) if you had as many coefficients as data points.

Alternative Hypothesis

The alternative hypothesis (written as H_a) is one of two choices that a hypothesis test must choose between; the other is the **null hypothesis** (written as H_0). The alternative hypothesis always involves a parameter in a strict inequality ($<$, $\neq$, or $>$). Here are three examples:

H_0	H_a
$\mu = 3$	$\mu \neq 3$
$\mu \geq 3$	$\mu < 3$
$\mu \leq 3$	$\mu > 3$

See **hypothesis testing**.

Analysis of Variance

Analysis of variance (ANOVA) is a hypothesis test that is used to test if several populations assumed to have a normal distribution with the same variance also have the same population means. It is possible to perform a **hypothesis test** using a **T random variable** if there are only two populations being compared (see **difference of means**); usually analysis of variance is only used for more than two populations. If μ_i represents the true (unknown) population mean for population i, then the null hypothesis states that the means for all k populations are the same:

$$\mu_1 = \mu_2 = \mu_3 = \ldots = \mu_k$$

The analysis of variance hypothesis test relies upon sample data from each population, and works best when each sample is of the same size.

The principle is the following: The sample means are computed and compared. If they are far apart from each other, that is evidence that the population means do differ. Their spread is measured by finding the variance of the sample means. It can be shown that if the null hypothesis is true then this variance $s_{\bar{x}}^2$ should be approximately σ^2/n, where σ^2 is the common population variance and n is the common sample size.

When is that variance too large? To get a yardstick, the individual sample variances are calculated and averaged together. Given that population variances are all equal, this average $\overline{s^2}$ should be approximately σ^2.

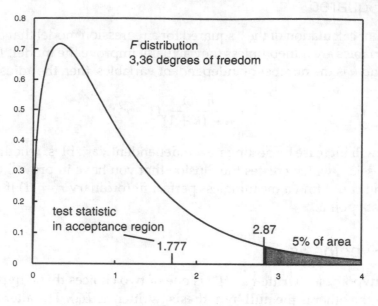

Figure 9.1

The ratio $\dfrac{ns_{\bar{x}}^2}{s^2}$ should be close to 1 if the null hypothesis is true, and larger than 1 if it is false. This ratio is an **F random variable** with $k - 1$ numerator degrees of freedom (where k is the number of populations) and with $k(n - 1)$ denominator degrees of freedom. Since the null hypothesis is disproved when this ratio is large, the test is a right-tailed test.

For example, suppose that we have the following sample statistics, each from a sample of size $n = 10$, for $k = 4$ different populations, all assumed to be normal with the same variance:

$$\bar{x}_1 = 23 \quad s_1^2 = 11$$
$$\bar{x}_2 = 25 \quad s_2^2 = 12$$
$$\bar{x}_3 = 22 \quad s_3^2 = 11$$
$$\bar{x}_3 = 22 \quad s_4^2 = 11$$

We want to test the null hypothesis that $\mu_1 = \mu_2 = \mu_3 = \mu_4$ at a significance level of $\alpha = 5\%$. Then $s_{\bar{x}}^2 = 2$, $\overline{s^2} = 11.25$, and $F = \frac{(10)(2)}{(11.25)} = 1.777$. For a right-tailed test with $4 - 1 = 3$ numerator degrees of freedom and $4(10 - 1) = 36$ denominator degrees of freedom, we have the critical value $F_{0.05} = 2.87$ (see page 211), and we accept the null hypothesis, since the evidence isn't strong enough to reject it (see Figure 9.1).

ANOVA

ANOVA is an abbreviation for **analysis of variance**.

Average

The average (symbolized by placing a bar over a quantity, as in $\bar{x}$) is a synonym for the **mean** of a group of n numbers $x_1, x_2, \ldots, x_n$, and is found by summing them and dividing by their quantity: $\bar{x} = \frac{(x_1 + x_2 + \ldots + x_n)}{n}$.

Bar Chart

A bar chart is a diagram where the height of different bars indicates the number of items in a particular category. For illustration, see **frequency histogram**.

Bayes' Rule

Bayes' rule is a formula relating the **conditional probabilities** of two events.

$$\Pr(A \mid B) = \frac{\Pr(B \mid A)\Pr(A)}{\Pr(B \mid A)\Pr(A) + \Pr(B \mid \text{not } A)\Pr(\text{not } A)}$$

For example, suppose that a test for a particular kind of cancer will detect the disease 95% of the time that it is present, and will give a false positive reading 4% of the time (i.e., will indicate cancer where there is none). Suppose further that 0.01% of the population actually has this cancer. If the test is positive, what is the probability that the patient actually has the cancer?

Let A be the event that the patient has the cancer; let B be the event that the test is positive. Then $\Pr(B \mid A) = .95$; $\Pr(B \mid \text{not } A) = .04$; and $\Pr(A) = .0001$. From Bayes' rule:

$$\Pr(A \mid B) = \frac{(0.95)(0.0001)}{(0.95)(0.0001) + (0.04)(0.9999)} = 0.00237 = 0.237\%$$

Therefore, the test is not a very good one—if the test is positive the chance is very small that the patient has the cancer.

Bayesian Approach to Decision Making

The Bayesian approach to decision making is an approach that uses what information is available, and incorporates new information as that, too, becomes available.

Bernoulli Trial

A Bernoulli trial is an experiment with only two possible outcomes, usually labeled success and failure. The probability p of success and probability q of failure are fixed, and must satisfy $p + q = 1$.

A **binomial random variable** counts the number of successes in n independent Bernoulli trials; a **geometric random variable** counts the number of independent Bernoulli trials until the first success.

Beta Random Variable

A beta random variable is a **continuous random variable** taking values between 0 and 1 that is often used as a model for proportions. It has two parameters α and β, and its probability density function is given by

$$f(x) = \begin{cases} \frac{\Gamma(\alpha+\beta)}{\Gamma(\alpha)\Gamma(\beta)} x^{\alpha-1}(1-x)^{\beta-1} & 0 < x < 1 \\ 0 & \text{otherwise} \end{cases}$$

Its mean and variance are given by:

$$\mu = \frac{\alpha}{\alpha + \beta}$$

$$\sigma^2 = \frac{\alpha\beta}{(\alpha + \beta)^2(\alpha + \beta + 1)}$$

The symbol Γ represents the **gamma function**. Figure 9.2 illustrates a sample density function.

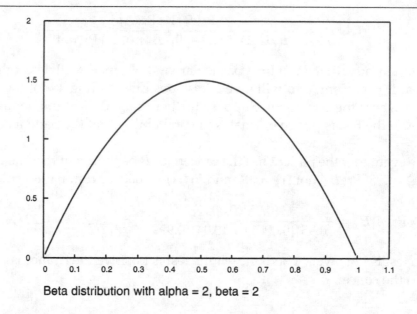

Beta distribution with alpha = 2, beta = 2

Figure 9.2: Probability density function for beta random variable

Biased Estimator

A biased estimator is an estimator whose average (i.e., expected) value is either larger or smaller than the parameter it is estimating. For contrast, see **unbiased estimator**.

Bimodal Distribution

A bimodal distribution refers to a population with two **modes**. A bimodal distribution often indicates the mixture of two separate populations. Figure 2.4 (page 13) illustrates the frequency distribution for a typical bimodal distribution.

Binomial Distribution

See **binomial random variable**.

Binomial Random Variable

The following experiment occurs frequently: A given trial with two possible outcomes labeled success and failure of fixed probabilities (labeled p and q, respectively, with $q = 1 - p$) is repeated independently a fixed number of times n; the number of successes X is a binomial random variable. The variable X can take whole values from 0 to n, and thus is a **discrete random variable**. Its **probability function** is given by

$$\Pr(X = k) = \begin{cases} \dbinom{n}{k} p^k q^{n-k} & 0 \le k \le n \\ 0 & \text{otherwise} \end{cases}$$

The expression $\dbinom{n}{k}$ represents the formula for **combinations**. The mean of X is np and its variance is $npq = np(1-p)$. The variance is zero if $p = 0$ or $p = 1$; for given n, the maximum variance occurs with $p = .5$. For example, if you toss a coin n times, the number of heads that appear is given by a binomial distribution with $p = 0.5$.

A binomial random variable can be thought of as a sum of n random variables R_i, where R_i is 1 if trial i is a success and 0 if it is a failure. Because it is the sum of a number of independent, identically distributed random variables, the binomial distribution can be approximated by a normal distribution for large values of n. This comes from the **Central Limit Theorem**.

Figure 9.3 illustrates a sample probability function for a binomial random variable.

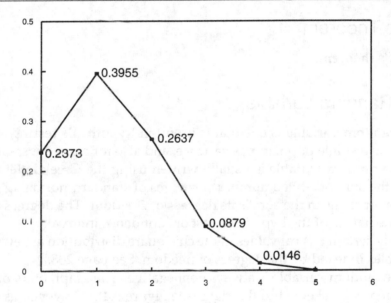

Figure 9.3: Binomial distribution with $n = 5, p = .25$.

Categorical Random Variable

A categorical random variable is a **random variable** that does not measure a number but rather a category. Examples include gender (with categories female and male) and automobile model. Categorical random variables are often coded numerically (e.g., 0 for female, 1 for male), but that is for convenience.

Central Limit Theorem

If a fixed **random variable** is measured repeatedly and independently, and the results averaged, the Central Limit Theorem claims that if the sample size is big enough, the resulting average will be a **normal random variable**. The Central Limit Theorem explains why normal random variables occur so frequently.

Formally, if X_1, X_2, X_3, $...X_n$ are independent random variables that all have the same distribution with mean μ and variance σ^2, then, as n becomes large, the random variable $\bar{x} = \frac{(X_1+X_2+X_3+...+X_n)}{n}$ will have a normal distribution with mean μ and variance σ^2/n.

If the distribution of X is itself normal, then $\bar{x}$ will automatically have a normal distribution because it is found by adding normal random variables and dividing by a constant. The power of the theorem is that it works regardless of the distribution of X, and it even works whether X is discrete or continuous.

For example, Figure 4.4 (page 52) illustrates the situation with dice. Even though the probabilities for one die do not look at all like a normal distribution, the probabilities for the average of three dice already look a little like a normal distribution, and the probabilities come closer and closer as the number of dice increases. Even though the formal theorem is written in terms of the limit as n approaches infinity, in practice the probabilities often start to look like a normal distribution even for relatively small values of n.

Chebyshev's Theorem

See **Tchebysheff's theorem**.

Chi-Square Random Variable

The chi-square random variable is one that is useful in **hypothesis testing** and **confidence intervals** involving a single population's variance, and also for a **goodness-of-fit test**.

The chi-square random variable is usually written using the Greek letter chi, i.e., as χ^2. Technically, it is the sum of a fixed number of squares of standard normal (Z) random variables; the number of terms in the sum is its **degrees of freedom**. The degrees of freedom are determined by the nature of the hypothesis test or confidence interval.

In practice, only certain critical values of the chi-square distribution are ever needed; they are read from tables indexed by the degrees of freedom. See page 208.

The chi-square random variable is always nonnegative. The graph of its probability density function is generally skewed to the right (see Figure 9.4). However, as the degrees of freedom becomes larger the chi-square distribution approaches the normal distribution. This follows from the **Central Limit Theorem**, since a chi-square random variable comes from the sum of independent, identically distributed random variables.

The probability density function for $x > 0$ is given by:

$$f(x) = \frac{x^{\frac{\nu}{2}-1}e^{\frac{-x}{2}}}{2^{\frac{\nu}{2}}\Gamma(\nu/2)}$$

Otherwise, $f(x) = 0$. In the formula ν is its degrees of freedom and Γ represents the **gamma function**.

The mean equals ν and the variance equals 2ν.

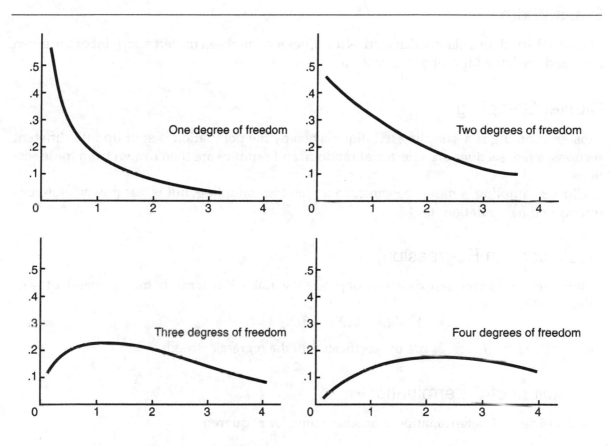

Figure 9.4: Probability density function for chi square random variable

Class

In the construction of a **frequency histogram**, numerical data is separated into intervals known as classes. The number of data points falling in each class can then be tabulated and used to describe the data.

Classes should be chosen as convenient intervals in the context of the given data, and should be of uniform width if at all possible, to make comparison between different sections of the data possible. The number of classes should usually fall between 5 and 15. Too few classes will over-simplify the description of the data, while too many classes will make the description hard to follow.

Class Width

The actual width of a class, calculated as the difference between the left endpoint of the given class and the left endpoint of the next class.

Cluster Sampling

Cluster sampling is a sampling technique whereby the population is split up into different sections, a few sections are selected at random, and samples are then drawn from these sections.

Cluster sampling is more economical than random sampling, but is less powerful in representing the population.

Coefficient (in Regression)

A linear **regression** model expresses a dependent variable Y in terms of the independent variables $X_1, X_2, \ldots, X_k$:
$$Y = \beta_0 + \beta_1 X_1 + \beta_2 X_2 + \ldots + \beta_k X_k$$
The numbers $\beta_0, \beta_1, \ldots, \beta_k$ are the coefficients of the regression model.

Coefficient of Determination

The coefficient of determination is another name for **r squared**.

Coefficient of Variation

The coefficient of variation is the ratio of the standard deviation to the mean, either for a sample ($s/\bar{x}$) or for a population (σ/μ). This ratio is usually expressed as a percentage. When the data refer to nonnegative measurements, the coefficient of variation indicates how significant the spread of data is as a fraction of the data's magnitude.

Combinations

The number of combinations is the number of ways of choosing k objects out of a group of n objects where two choices are considered to be the same if they contain the same k objects, without regard to their order of selection. The number of ways of choosing is given by $n!/[k!(n-k)!]$, which is symbolized $\binom{n}{k}$ or C_k^n:

$$\binom{n}{k} = C_k^n = \frac{n!}{k!(n-k)!}$$

To derive this formula, note there are n choices for the first item chosen, $n-1$ choices for the second one, and so on down to $n-k+1$ choices for the kth item chosen. Multiply these together to give the total number of ways of choosing the objects:

$$n \times (n-1) \times (n-2) \times \ldots \times (n-k+1) = \frac{n!}{(n-k)!}$$

The exclamation point designates *factorial*. However, the above formula counts each ordering of the objects separately (this is known as the number of **permutations**). We need to divide by the $k!$ different orderings to come up with the formula for combinations:

$$\binom{n}{k} = \frac{\frac{n!}{(n-k)!}}{k!} = \frac{n!}{k!(n-k)!}$$

Some special properties are:

$$\binom{n}{0} = \binom{n}{n} = 1$$

$$\binom{n}{1} = \binom{n}{n-1} = n$$

$$\binom{n}{k} = \binom{n}{n-k}$$

The formula for combinations is used in the definition of the **binomial random variable** and the **hypergeometric random variable**.

Conditional Probability

Given two events A and B, the conditional probability of A given B is the probability that A will occur, given that B has occurred. If B has nonzero probability, then the conditional probability of A given B is

$$\Pr(A|B) = \frac{\Pr(A \text{ and } B)}{\Pr(B)}$$

If A and B are **independent events**, then the probability of A occurring given that B has occurred is just the probability of A.

If A and B are **mutually exclusive** events, then $\Pr(A \mid B) = 0$.

See also **Bayes' rule**.

Confidence Interval

A major component of inferential statistics consists of the estimation of the parameter(s) of a population. In certain circumstances a single number, i.e., an estimator, is appropriate; more often a confidence interval is more useful.

A confidence interval is an interval (with two endpoints that may be infinite) that will contain the given parameter with a predetermined probability. The endpoints (and thus the interval) are determined from sample data and thus are themselves random variables.

This is an important point. The confidence interval endpoints are themselves random variables, which will vary according to the sample data. The parameter that we hope the confidence interval will contain is a fixed (non-random) but unknown quantity.

The probability that a given confidence interval will contain the parameter is called the confidence level. It is usually set at either 95% or 99%, and is used in determining the endpoints of the interval. Thus if the confidence level were set at 99% and you constructed 100 confidence intervals, then you would expect roughly 99 of them to contain the parameter that you sought, and 1 of them to miss it. The larger the confidence level, the larger the width of the confidence interval.

Theoretically, there are many ways of constructing confidence intervals; in practice there are a few fixed ways of doing so for estimating the most common population parameters. Three common examples are included here.

Confidence Interval for Population Mean

A confidence interval for the population mean will use the standard normal (Z) random variable or the T random variable depending upon the sample size. The formula for the endpoints of the interval is:

$$\bar{x} \pm \frac{as}{\sqrt{n}}$$

Here $\bar{x}$ is the sample average; s is the sample standard deviation; and n is the sample size. If the sample size is larger than about 30, then a comes from the standard normal (Z) table:

$$\Pr(-a < Z < a) = CL$$

The quantity CL is the confidence level. If the confidence level is .95, then $a = 1.96$. This table shows the value of a for other confidence levels:

Confidence level (CL)	a (chosen so $\Pr(-a < Z < a) = CL$)
.80	1.28
.85	1.44
.90	1.65
.95	1.96
.99	2.58

If the sample size is smaller than 30, then a comes from the two-tail t distribution table (page 209) for $n-1$ degrees of freedom:

$$\Pr(-a < T < a) = CL$$

Confidence Interval for Proportion

If p is the unknown proportion of type A objects in the population, and a random sample of size n is selected from this population with $\hat{p}$ being the proportion of type A objects in the sample, then a 95% confidence interval for p is found to be the interval from

$$\hat{p} - 1.96\sqrt{\frac{\hat{p}(1-\hat{p})}{n}} \quad \text{to} \quad \hat{p} + 1.96\sqrt{\frac{\hat{p}(1-\hat{p})}{n}}$$

For confidence levels other than 95%, change the value 1.96 to the value given in the previous table.

Confidence Interval for Variance

To find a 95% confidence interval for the variance of a sample of size n chosen from an approximately normal population, first find values a and b such that

$$\Pr(\chi^2_{n-1} < a) = .025 \quad \text{and} \quad \Pr(\chi^2_{n-1} < b) = .975$$

The random variable χ^2_{n-1} has a chi-square distribution with $n-1$ degrees of freedom. For example, if $n = 25$, then there are 24 degrees of freedom and we see from the chi-square table (page 208) that $a = 12.40$ and $b = 39.4$.

Calculate $\hat{s}^2$, which is the variance using n instead of $n-1$ in the denominator:

$$\hat{s}^2 = \frac{(X_1 - \bar{x})^2 + \ldots + (X_n - \bar{x})^2}{n} = \overline{x^2} - \bar{x}^2$$

Then the confidence interval is from:

$$\frac{n\hat{s}^2}{a} \quad \text{to} \quad \frac{n\hat{s}^2}{b}$$

For other confidence levels (CL), find a and b from:

$$\Pr(\chi^2_{n-1} < a) = \frac{1-CL}{2} \quad \text{and} \quad \Pr(\chi^2_{n-1} < b) = \frac{1+CL}{2}$$

See also **difference of means**.

Confidence Level

The confidence level is the probability that a given **confidence interval** will contain the population parameter that it is intended to estimate. The confidence level is determined prior to constructing the confidence interval and is used in that construction.

The most commonly used confidence levels are 95% and 99%. The larger the confidence level, the wider the confidence interval.

Consistent Estimator

A consistent **estimator** $\hat{\theta}_n$ for a population parameter θ is one whose variance approaches 0 as the sample size n approaches infinity.

Continuous Random Variable

A continuous random variable is one whose possible values lie in an interval (and thus are not isolated points). Examples include the **normal random variable**, **T random variable**, **chi-square random variable**, and **F random variable**.

There is zero probability that a continuous random variable X will exactly hit any one specific value. The probability that X will be between two values a and b is given by the area under its **density function** between those values (see Figure 4.6, page 58).

This area can be expressed as an **integral**:

$$\Pr(a < X < b) = \int_a^b f(x)dx$$

The function $f(x)$ is the density function. The total area under the density function must be 1:

$$\int_{-\infty}^{\infty} f(x)dx = 1$$

The expected value is defined by this integral:

$$E(X) = \int_{-\infty}^{\infty} xf(x)dx$$

Unfortunately, there are no formulas for most of the areas needed in statistics, so it is necessary to consult a table or rely on a computer program's built-in function to find the areas.

For contrast, see **discrete random variable**.

Correlation Coefficient

Given two numerical random variables defined on the same data points, the correlation coefficient measures the extent to which their relationship is a straight line on the scatterplot.

The correlation coefficient is a number between -1 and 1. A value of -1 indicates that the points on the **scatterplot** lie on a straight line of negative slope; a value of 1 indicates that those points lie on a straight line of positive slope.

The correlation coefficient for sample data is often symbolized by r. Its square is the r^2 value used in simple regression analysis.

The formula for the correlation is:

$$r = \frac{\overline{xy} - \overline{x}\,\overline{y}}{\sqrt{(\overline{x^2} - \overline{x}^2)(\overline{y^2} - \overline{y}^2)}} = \frac{\overline{xy} - \overline{x}\,\overline{y}}{\sqrt{\mathrm{Var}(X)\,\mathrm{Var}(Y)}}$$

The correlation coefficient for population data is often symbolized by ρ (rho).

Here is a simple example of the calculation process:

	X	Y	XY	X^2	Y^2
	4	10	40	16	100
	7	8	56	49	64
	3	1	3	9	1
	6	9	54	36	81
Average:	5	7	38.25	27.5	61.5

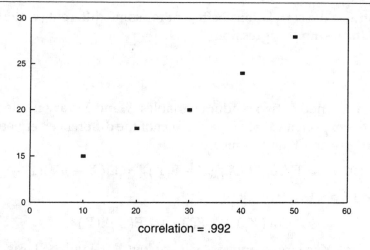

correlation = .992

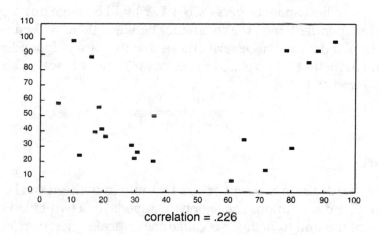

correlation = .226

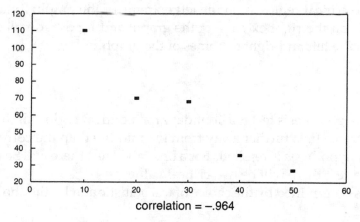

correlation = −.964

Figure 9.5

$$r = \frac{38.25 - 5 \times 7}{\sqrt{(27.5 - 5^2)(61.5 - 7^2)}} = .581$$

Figure 9.5 illustrates the correlation coefficient for some different scatterplots. See also **covariance**, **simple regression**.

Covariance

Covariance measures whether two random variables X and Y vary in the same way or not. It is defined to be the expected value of the product of the difference between X and its mean and the difference between Y and its mean:

$$\text{Cov}(X, Y) = \text{E}[(X - \text{E}(X))(Y - \text{E}(Y))] = \text{E}[(X - \mu_X)(Y - \mu_Y)]$$

It can also be found from this formula:

$$\text{Cov}(X, Y) = \text{E}(XY) - \text{E}(X)\text{E}(Y)$$

If X is largest when Y is largest and vice versa, $\text{Cov}(X, Y)$ will be large and positive; if X is largest when Y is smallest and vice versa, $\text{Cov}(X, Y)$ will be large but negative.

If X and Y are independent, then the covariance between them will be zero. However, if the covariance is zero, it does not necessarily mean that they are independent.

The **correlation coefficient** (r) is related to the covariance; it is scaled so that its value is always between -1 and 1:

$$r = \frac{\text{Cov}(X, Y)}{\sqrt{\text{Var}(X)\,\text{Var}(Y)}}$$

Critical Region

In a hypothesis test, the decision to accept or reject the null hypothesis can be indicated on the graph of the probability distribution. The region (or regions for a two-tailed test) corresponding to the rejection of the null hypothesis is called the critical region or rejection region. For a left-tailed test, the critical region is on the left extreme of the graph; for a right-tailed test, the critical region is on the right extreme of the graph; and for a two-tailed test, the critical regions are on both the left and right extremes of the graph.

Critical Value

A critical value for a hypothesis test is a boundary for a **critical region**. If the value of the test statistic is more extreme (i.e., further away from the middle) than the corresponding critical value, then the null hypothesis is rejected. For a one-tailed test there will be one critical value, and for a two-tailed test there will be two critical values.

A critical value is determined by the choice of test statistic and by the choice of significance level.

Cross Tabulation

When data can be categorized along two dimensions (e.g., eye color versus hair color), a listing of the frequencies for each combination of the two dimensions (e.g., how many in the sample have green eyes and brown hair, etc.) is known as a cross tabulation of the data.

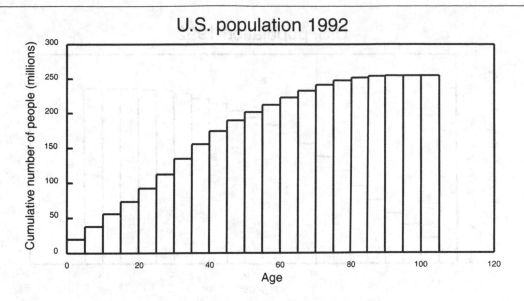

Figure 9.6: Cumulative frequency histogram

Cumulative Frequency

The running total of the number of data points in a given class or any of its preceding classes.

Cumulative Frequency Histogram

A histogram similar to a frequency histogram. Numerical data is split into classes, and the frequencies for each class calculated. Cumulative frequencies are then calculated and plotted as a bar chart, with the width of each bar corresponding to the class width (see Figure 9.6).

Cumulative Relative Frequency

The cumulative relative frequency for a given class is the fraction of the number of data points that are contained in the given class and all preceding classes. It is calculated by taking the cumulative frequency for the given class and dividing by the number of data points. Cumulative relative frequencies are usually expressed as percentages.

Cumulative Relative Frequency Histogram

A histogram similar to a frequency histogram. Numerical data are split into classes, and the frequencies for each class calculated. Cumulative frequencies and then cumulative relative frequencies are calculated, and the cumulative relative frequencies are plotted as a bar chart, with the width of each bar corresponding to the class width (see Figure 9.7).

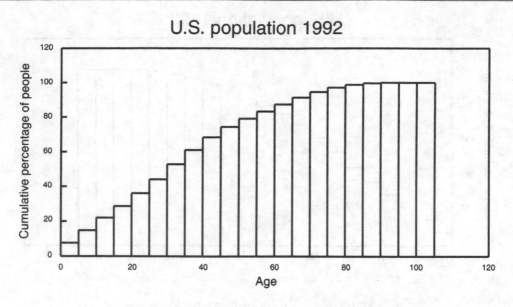

Figure 9.7: Cumulative relative frequency histogram

Data Points

A data point refers to a discrete unit in a collection of data. For example, in a survey of adult human weights and heights, each person in the survey would be a data point, and each data point would have two numbers attached to it, namely, the person's weight and height.

Deciles

If a collection of numbers is put in order and split into ten parts, the boundary points are the deciles. The deciles correspond to the 10th, 20th, 30th, 40th, 50th, 60th, 70th, 80th, and 90th **percentiles**.

Decision Theory

Decision theory is the study of how to make advantageous decisions on the basis of uncertain information.

Degrees of Freedom

Many random variables, such as the **chi-square random variable** and the **T random variable**, have associated with them a whole number known as their degrees of freedom. This phrase has both a technical and an intuitive meaning.

Technically, the chi-square random variable is defined to be the sum of the squares of a fixed number of standard normal (Z) random variables; the number of terms in the sum is the degrees of freedom for that chi-square random variable. The T random variable is defined to be the ratio of a Z random variable to the square root of a chi-square random variable; the T random variable is said to have the same degrees of freedom as that chi-square random variable. The F random variable is defined to be proportional to the ratio of two chi-square random variables; correspondingly, the F random variable has numerator degrees of freedom and denominator degrees of freedom.

There isn't a need in practice to go into these technical details. Intuitively, the degrees of freedom refers to the number of sample points that remain independent. For example, if in the calculation of a T random variable based on a sample of size 20, the sample mean is used, then the T random variable has 19 degrees of freedom. The 20 sample points have one constraint placed upon them once we have calculated the sample mean; 19 of the sample points could take any value, but then the twentieth would be determined by the sample mean.

In each context involving the chi-square random variable, the F random variable, and the T random variable, there will be a fixed rule for calculating the degrees of freedom, and this rule will always yield the number of sample points that remain independent.

Denominator Degrees of Freedom

The **F random variable** has two different **degrees of freedom**: one for the numerator and the other for the denominator. This occurs because formally the F random variable is defined to be proportional to the ratio of two chi-square random variables, each with its own degrees of freedom. In practice, the context will determine both the numerator degrees of freedom and the denominator degrees of freedom for an F random variable.

Density Function

A **continuous random variable** X has associated with it a density function $f(x)$ such that

$$\Pr(X < a) = \int_{-\infty}^{a} f(x)dx$$

and

$$E[g(X)] = \int_{-\infty}^{\infty} g(x)f(x)dx$$

See **integral**.

Dependent

Two events or random variables are dependent if they are not independent. See **independence**.

Dependent Variable

In a **regression** model, the output variable is the dependent variable. For example, Y is the dependent variable in the regression equation

$$Y = B_0 + B_1 X_1 + B_2 X_2 + \epsilon$$

For contrast, see **independent variable**.

Descriptive Statistics

Descriptive statistics are numbers or diagrams that are calculated to describe a given collection of data, usually a sample. There are five main kinds of descriptive statistics that are commonly used:

- Descriptions of where the center of the data lies. "Center" is an ambiguous term, and thus there is more than one way to describe the center. The most common ones are the **mean**, the **median**, and the **mode**; each has its own strengths and weaknesses.

- Descriptions of how the data are spread out. These include the **range**, the **variance**, the **standard deviation**, and the **interquartile range**.

- Descriptions of the relative positions of the data, including **percentiles** and the related **quartiles** and **deciles**.

- A **frequency histogram** is a bar chart that displays how the data is spread out. Common variations include a **relative frequency histogram**, a **cumulative frequency histogram**, and a **cumulative relative frequency histogram**.

- If the sample data include more than one kind of measurement (e.g., weight and height), then the relationship between two kinds of measurements can be shown on a **scatterplot**.

Difference of Means

One of the most frequent questions in statistics is whether or not two populations possess the same mean. If we assume that the populations each have a normal distribution, then it is possible to construct a **confidence interval** for the difference of the population means, centered around the difference of the sample means. It is also possible to perform a **hypothesis test** as to whether or not the means are equal.

To construct a confidence interval for the difference of the population means μ_1 and μ_2, we start with the difference of the sample means $\bar{x}_1$ and $\bar{x}_2$. The difference $\bar{x}_1 - \bar{x}_2$ is itself a random variable, with the following variance:

$$\sigma^2_{\bar{x}_1-\bar{x}_2} = \frac{\sigma_1^2}{n_1} + \frac{\sigma_2^2}{n_2}$$

The quantities σ_1^2 and σ_2^2 are the population variances, and n_1 and n_2 are the two sample sizes.

If the sample sizes are both more than 30, we can estimate the population variances using the **sample variances** s_1^2 and s_2^2 to construct a confidence interval using the standard normal (Z) random variable, namely:

$$\left(\bar{x}_1 - \bar{x}_2 - a\sqrt{\frac{s_1^2}{n_1} + \frac{s_2^2}{n_2}} \quad \text{to} \quad \bar{x}_1 - \bar{x}_2 + a\sqrt{\frac{s_1^2}{n_1} + \frac{s_2^2}{n_2}} \right)$$

If the 95% confidence level is used, $a = 1.96$. For other confidence levels, the value of a comes from the standard normal table (page 205):

$$\Pr(-a < Z < a) = CL$$

where CL is the confidence level.

If either of the **sample sizes** are smaller than 30, then we must use a similar formula based on the **T random variable**, namely

$$\left(\overline{x}_1 - \overline{x}_2 - a\sqrt{\frac{s_1^2}{n_1} + \frac{s_2^2}{n_2}} \quad \text{to} \quad \overline{x}_1 - \overline{x}_2 + a\sqrt{\frac{s_1^2}{n_1} + \frac{s_2^2}{n_2}} \right)$$

Here a comes from the two-tail t table (page 209):

$$\Pr(-a < T < a) = CL$$

Again, CL is the confidence level; T is a random variable with the t distribution with degrees of freedom equal to the minimum of $n_1 - 1$ and $n_2 - 1$.

To test the hypothesis that $\mu_1 = \mu_2$ for large sample sizes we can use the standard normal test statistic:

$$Z = \frac{\overline{x}_1 - \overline{x}_2}{\sqrt{\frac{s_1^2}{n_1} + \frac{s_2^2}{n_2}}}$$

If the sample sizes are small, we will use the following test statistic:

$$T = \frac{\overline{x}_1 - \overline{x}_2}{\sqrt{\frac{s_1^2}{n_1} + \frac{s_2^2}{n_2}}}$$

This statistic has a t distribution with degrees of freedom equal to the minimum of $n_1 - 1$ and $n_2 - 1$ if the null hypothesis is true.

Discrete Random Variable

A discrete random variable is one that can only take isolated values. For example, if a random variable can only take on whole number values, it is a discrete random variable; its possible values are separated from each other. Examples include the **binomial random variable** and the **geometric random variable**. By contrast, a **continuous random variable** can take on any value within an interval.

Disjoint

Two events A and B are disjoint if they can never happen simultaneously, or $\Pr(A \text{ and } B) = 0$. This is a synonym for **mutually exclusive**.

Dummy Variable

In a **regression** model it is often desirable to include **categorical random variables** along with numerical variables as **independent variables**. The different categories can each be assigned a separate dummy variable that will be 1 for those data points in a given category and 0 otherwise.

Durbin-Watson

In a time-dependent **regression** model, the **residuals** are often **dependent** on each other. The Durbin-Watson test statistic

$$d = \frac{\sum_{t=2}^{n}(\hat{e}_t - \hat{e}_{t-1})^2}{\sum_{t=1}^{n}\hat{e}_t^2}$$

(where $\hat{e}_t$ is the residual term at time t) provides a way of testing whether the correlation between successive error terms is zero, positive, or negative. A small value of the statistic indicates the presence of serial correlation; the observed value needs to be compared with values given in a special table.

e

The letter e represents a number approximately equal to 2.71828. It is the base of the natural **logarithm** function, and it is used in the definition of the density function for a **normal random variable**, among others.

Error

Error is the difference between the true value of a dependent variable and the corresponding prediction of a regression model.

Estimator

Rarely can we know precisely the parameters connected with a given population; the amount of time, money, and energy needed to find the parameters is just too great.

Instead, when we want to find a parameter for a given population, we usually settle on finding an estimate based on sample data. A formula or method for finding an estimate is called an estimator.

For example, the sample mean $\bar{x}$ is an estimator for the population mean μ.

Sometimes a hat is placed over a sample statistic that is used as an estimator; for example, $\hat{\theta}$ would designate an estimator for a parameter θ.

See also **unbiased estimator**, **consistent estimator**, **maximum likelihood estimator**.

Event

An event is a collection of possible outcomes for an experiment. For example, if the experiment consisted of rolling two six-sided dice and noting the numbers on the top of the two dice, the event "the sum of the dice is 7" consists of the outcomes (1, 6), (2, 5), (3, 4), (4, 3), (5, 2), and (6, 1).

Expectation

The expectation of a random variable is the same as its **expected value**.

Expected Value

The expected value of a random variable X is the average value that would appear if this random variable were observed many times. It is symbolized by $E(X)$ or μ, and is also called the mean.

More generally, the expected value (or expectation) of a function $g(X)$ of a **random variable** X is the average value $E[g(X)]$ that $g(X)$ takes as X runs through all of its possible values. If X is a **discrete random variable**, the averaging is done using the probabilities at each possible value:

$$E[g(X)] = \sum g(a)\Pr(X = a)$$

The summation is done as a runs through all possible values. If X is a **continuous random variable**, then the averaging is done using X's **density function** $f(x)$:

$$E[g(X)] = \int_{-\infty}^{\infty} g(x)f(x)dx$$

The **variance** is defined as the expected value of the squared distance from the mean:

$$\mathrm{Var}(X) = \sigma^2 = E[(X - \mu)^2]$$

This can also be expressed as:

$$\mathrm{Var}(X) = \sigma^2 = E(X^2) - [E(X)]^2$$

The expected value satisfies these properties:

- $E(cX) = cE(X)$,
  where c is any constant and X is any random variable.

- $E(X + Y) = E(X) + E(Y)$,
  where X and Y are any two random variables.

- $E(XY) = E(X)E(Y)$,
  where X and Y are two independent random variables.

Experiment

An experiment is a procedure that determines an **outcome**. Prior to the experiment there should be a fixed list of possible outcomes but no foreknowledge as to which will occur. After the experiment there should be complete certainty as to which outcome has occurred. For example, tossing a die or choosing a random sample are experiments in statistics.

Exponential Random Variable

The exponential random variable is a continuous analog of the **geometric random variable** in that it can only take positive values and that the probabilities associated with it decline at a constant rate. An exponential random variable X is memory-less in the following way: for a and b positive, $\Pr(X > a + b | X > b) = \Pr(X > a)$. To put it another way, if X is the time it takes for a light bulb to burn out, then X does not depend on how long the bulb has already

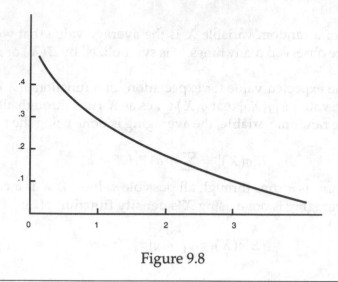

Figure 9.8

been burning. The exponential random variable is determined by one parameter β, and its probability density function is given by

$$f(x) = \begin{cases} \frac{1}{\beta}e^{-x/\beta} & x > 0 \\ 0 & \text{otherwise} \end{cases}$$

Its mean is β and its variance is β^2.

Figure 9.8 illustrates a sample density function for an exponential distribution.

Extrapolation

Extrapolation refers to the process of using a **regression** model to make a prediction based on values for the independent variable(s) that lie outside those of the sample data. Extrapolation is always riskier than interpolation, because the pattern found in the sample data may change outside its boundary. For illustration, see Example 11, page 135.

F Distribution

See **F random variable**.

F Random Variable

An F random variable is a random variable that is used in hypothesis tests about variances for two populations, **analysis of variance**, and decisions about which variables to include in **regression** models.

Formally, an F random variable is defined to be

$$\frac{\chi_1^2/\nu_1}{\chi_2^2/\nu_2}$$

Here χ_1^2 is a chi-square random variable with ν_1 degrees of freedom, and χ_2^2 is a chi-square random variable with ν_2 degrees of freedom.

In order to identify a specific F random variable, it is necessary to specify both the numerator degrees of freedom (ν_1 in the above example) and the denominator degrees of freedom (ν_2). An F random variable is always nonnegative.

Since only certain critical values for an F random variable are ever needed, they are indexed in tables according to the numerator and denominator degrees of freedom (see page 211). Figure 9.1 (page 142) illustrates a sample density function for an F random variable.

F Statistic in Regression

In a **regression** model it is possible not only to test one coefficient at a time to see if it is zero, but also to test a whole group of coefficients simultaneously to see if they are all zero.

Suppose that you have a least squares linear regression model $Y = \beta_0 + \beta_1 X_1 + \ldots + \beta_{m-1} X_{m-1} + \epsilon$, where ϵ is assumed to be normal with mean 0 and variance σ^2. If you want to test the null hypothesis $H_0 : \beta_{g+1} = \beta_{g+2} = \ldots = \beta_k = 0$, it is natural to look at the difference between the sum of squared error SSE_1 coming from the small model $Y = \beta_0 + \beta_1 X_1 + \ldots + \beta_g X_g$ and that coming from the original, bigger model, SSE_2. If SSE_1 is significantly bigger than SSE_2, then that is evidence that the independent variables $X_{g+1}, X_{g+2}, \ldots, X_{m-1}$ really do belong in the model and that the null hypothesis should be rejected.

We will use an F random variable for our **test statistic**. Let n be the sample size. Then there are two estimators for σ^2:

$$S_{dif}^2 = \frac{SSE_1 - SSE_2}{m - 1 - g}$$

and

$$S_{big}^2 = \frac{SSE_2}{n - m}$$

The ratio of the estimators $F = S_{dif}^2 / S_{big}^2$ is an F random variable with $m - 1 - g$ numerator degrees of freedom and $n - m$ denominator degrees of freedom. Since H_0 is rejected when $SSE_1 - SSE_2$ is large, i.e., when F is large, this is a right-tailed test.

A computer regression program will typically report the value of the F statistic that tests the hypothesis that all coefficients are zero. This F statistic will come from an F distribution with $m - 1$ numerator degrees of freedom and $n - m$ denominator degrees of freedom if the null hypothesis is true.

Factorial

If n is a positive integer, then its factorial is given by

$$n! = (n) \times (n - 1) \times (n - 2) \times \ldots \times (2) \times (1)$$

For example, $4! = 4 \times 3 \times 2 \times 1 = 24$. There are $n!$ different ways of putting n objects in order. For applications, see **combinations** and **permutations**.

Finite Population Correction Factor

When a sample of size n is taken without replacement from a large population of size N, it is nearly the same as **sampling with replacement**. When a sample is more than 5% of a population, it is necessary to introduce a finite population correction factor into the calculation of the standard deviation of the sample mean, i.e., replace $\sigma_{\bar{x}} = \sigma/\sqrt{n}$ by

$$\sigma_{\bar{x}}\sqrt{\frac{N-n}{N-1}} = \frac{\sigma}{\sqrt{n}}\left(\sqrt{\frac{N-n}{N-1}}\right)$$

To see where this comes from, see **hypergeometric random variable**.

Frequency

For a given class, the frequency is the number of data points contained in the class.

Frequency Histogram

A frequency histogram is a bar chart where the widths of the bars correspond to the classes for the data points, and the heights of the bars correspond to the frequencies for the classes.

For example, here is a table giving the population of the U.S. in 1992, divided into age classes each 5 years wide (numbers are in thousands):

Age range	Frequency
0 to 5	19,512
5 to 10	18,349
10 to 15	18,099
15 to 20	17,074
20 to 25	19,051
25 to 30	20,190
30 to 35	22,271
35 to 40	21,098
40 to 45	18,806
45 to 50	15,361
50 to 55	12,056
55 to 60	10,487
60 to 65	10,440
65 to 70	9,979
70 to 75	8,482
75 to 80	6,415
80 to 85	4,152
85 to 90	2,162
90 to 95	834
95 to 100	218
over 100	45
Total	255,081

Figure 9.9 illustrates the frequency histogram for this data.

Normally, a frequency histogram will be drawn with classes of equal width. However, in some cases it will be necessary to use unequal widths, in which case the height of each bar is proportional to the frequency of the class divided by the width of the interval. This makes sure that the wider classes do not have misleadingly high bars.

See also **cumulative frequency histogram**, **relative frequency histogram**, **cumulative relative frequency histogram**, and **ogive**.

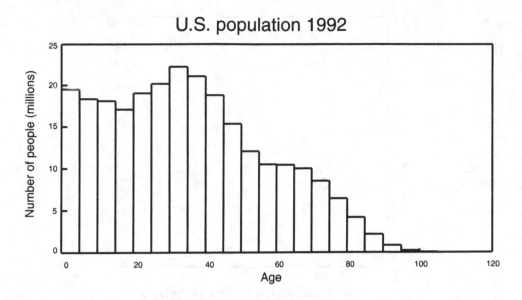

Figure 9.9: Frequency histogram

Gamma Function

The gamma function, usually denoted $\Gamma(x)$, is a function that generalizes the **factorial function**. The connection between the two is that $\Gamma(n + 1) = n!$ when n is a nonnegative integer. The gamma function, however, is defined for fractional numbers as well as for the positive integers; its formula is given by

$$\Gamma(x) = \int_0^\infty t^{x-1}e^{-x}dx$$

Three general properties are:

- $\Gamma(0) = 1$

- $\Gamma(1/2) = \sqrt{\pi}$

- $\Gamma(n) = (n - 1)\Gamma(n - 1)$

Gamma Random Variable

A gamma random variable is a **continuous random variable** X that only takes nonnegative values and whose probability distribution is skewed to the right. It is controlled by two positive parameters α and β, with probability density function given by

$$f(x) = \frac{\beta^\alpha}{\Gamma(\alpha)}x^{\alpha-1}e^{-bx}$$

Its mean and variance are given by $\mu = \alpha/\beta$ and $\sigma^2 = \alpha/\beta^2$.
Figure 9.10 illustrates a sample probability function for a gamma random variable.

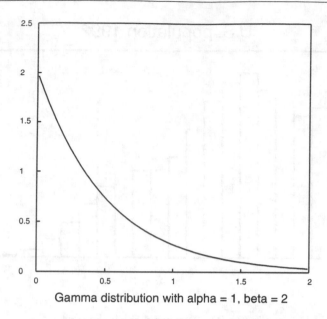

Figure 9.10: Probability density function for gamma random variable

Geometric Mean

The geometric mean of two positive numbers a and b is the square root of their product, i.e., $\sqrt{ab}$.

Geometric Random Variable

Consider the following open-ended experiment: trials with two possible outcomes (labeled success and failure) with fixed probabilities (p and q respectively, with $q = 1 - p$) are repeated independently until the first success. The number of trials X until the first success is called a geometric random variable; it can take values of 1, 2, 3, ... The graph of its probability distribution is skewed to the right. The probability of $X = k$ for $k \geq 1$ is pq^{k-1}. Its mean is $1/p$ and its variance is q/p^2. See also **binomial random variable** and **negative binomial random variable**.

Figure 9.11 illustrates a sample probability function for a geometric random variable.

Goodness-of-Fit Test

A goodness-of-fit test is a hypothesis test to decide if a given model accurately describes a given population. There are two commonly used versions of the goodness-of-fit test:

- There is a one-dimensional version, where the model for the data is a **multinomial random variable**. Sample data are collected, and the number of data points falling into each category are compared with those predicted by the multinomial distribution. In this case the degrees of freedom are one less than the number of categories.

- There is a two-dimensional version, which tests if two categorical random variables are independent. The proportions of the data points in each of the categories of the two

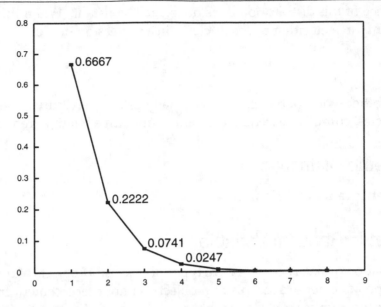

Figure 9.11: Geometric distribution with p = $\frac{2}{3}$

random variables are calculated separately, and then multiplied to find the proportion of data points to be expected (if the two random variables were really independent) in each joint category. As with the one-dimensional version, the observed and expected frequencies are compared. In this case the degrees of freedom are the product $(r-1)(c-1)$, where r is the number of categories for the first random variable (listed as rows), and c is the number of categories for the second random variable (listed as columns).

In each of the two cases, the test variable is a **chi-square random variable**, where

$$\chi^2 = \sum_i \frac{(O_i - E_i)^2}{E_i}$$

The variable O_i is the number of observations in category #i, and E_i is the expected (possibly fractional) number of observations in category #i. This hypothesis test is always a right-tailed test, since a large value for χ^2 indicates strong disagreement between the observed and the expected values.

Harmonic Mean

The harmonic mean of two positive numbers a and b is the reciprocal of the average of their reciprocals, i.e.,

$$\frac{1}{\frac{\frac{1}{a}+\frac{1}{b}}{2}} = \frac{2ab}{a+b}$$

Heteroscedasticity

For a given set of values for the independent variables in a **regression** model there is a probability distribution for the true values of the independent variables. Heteroscedasticity occurs when the variance of this distribution does not remain constant. When this applies there is greater uncertainty in estimating the coefficients in a regression model.

Histogram

Histogram is another word for bar chart. It is usually drawn vertically to depict frequencies, relative frequencies, cumulative frequencies, and cumulative relative frequencies.

Hypergeometric Distribution

See **hypergeometric random variable.**

Hypergeometric Random Variable

Suppose a population of size N contains two types of objects: M objects of type M and $N - M$ objects not of type M. Choose a sample of n objects without replacement from this population, and let X be the number of objects in the sample of type M. Then X is a hypergeometric random variable. It is a **discrete random variable** that can take values from 0 to the minimum of M and n. Its probability function is given by

$$
\Pr(X = k) = \begin{cases} \dfrac{\dbinom{M}{k}\dbinom{N-M}{n-k}}{\dbinom{N}{n}} & 0 \leq k \leq \min(M, n) \\[2em] 0 & \text{otherwise} \end{cases}
$$

where

$$
\binom{N}{n}
$$

is the notation for the number of **combinations** of n items selected from a population of N items. To derive this formula, note that there are

$$
\binom{N}{n}
$$

ways to select the sample; this appears in the denominator because it is the total number of possible outcomes. There are

$$
\binom{M}{k}
$$

ways of selecting k objects from the M objects of type M; and there are

$$
\binom{N-M}{n-k}
$$

ways of selecting $n - k$ non-type M objects from among the $N - M$ non-type M objects in the population. These two numbers need to be multiplied to give the total number of outcomes that contain k objects of type M and $n - k$ non-type M objects; this product appears in the numerator of the hypergeometric probability formula.

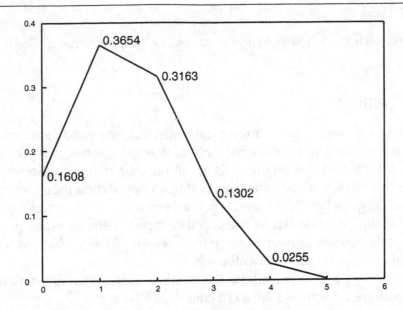

Figure 9.12: Hypergeometric distribution: $N = 100$, $M = 30$, $n = 5$

The mean of a hypergeometric random variable is nM/N. Intuitively this makes sense; if $M = 600$ and $N = 1,000$, then you would expect 60% of the objects in the sample will be of type M. If $n = 10$, then the expected value of X would be 6.

The variance is given by:

$$n\left(\frac{M}{N}\right)\left(1 - \frac{M}{N}\right)\left(\frac{N-n}{N-1}\right)$$

If we let $p = M/N$ be the proportion of type M items in the population, then the variance can be written:

$$np(1-p)\left(\frac{N-n}{N-1}\right)$$

Note that if N is much larger than n, then the factor in parentheses on the right becomes close to 1, in which case the variance becomes the same as the variance of a **binomial random variable** with probability of success p. The binomial distribution would apply if you sampled n times with replacement from a population where p gives the proportion of objects of type M. The hypergeometric distribution applies if you sample without replacement, but if the population is much larger than the sample it doesn't make much difference whether the sample is chosen with or without replacement. This makes it possible to use the binomial distribution (which is easier to deal with) as an approximation for the hypergeometric distribution. See also **finite population correction factor**.

Figure 9.12 illustrates a sample probability function for a hypergeometric random variable.

Hypothesis

A hypothesis is an assertion that can be tested by collecting and analyzing data. See **hypothesis testing**.

Hypothesis Test

A procedure to determine whether to accept or reject a null hypothesis. See **hypothesis testing**.

Hypothesis Testing

Hypothesis testing is a major part of **inferential statistics**. Any good experiment is designed to collect data that will test whether or not a given statement is true. Hypothesis testing is a formal procedure to do just that on the basis of sample data that are collected.

A hypothesis test starts with two statements about a population that are opposites of each other, e.g., the average height of all men is 5'7"versus the average height of all men is not 5'7". Typically the statements will refer to a population parameter such as a population mean. Sometimes the statements will refer to more than one population, such as a claim that the means of five different populations are all equal.

Since the population parameter (call it θ) is a number, these statements will have one of the following three different forms, where a is a constant:

- $\theta = a$ versus $\theta \neq a$

- $\theta \geq a$ versus $\theta < a$

- $\theta \leq a$ versus $\theta > a$

Traditionally, the hypothesis that includes equality is called the null hypothesis, written H_0. The first statement in each of the three forms listed above is a **null hypothesis**. The hypothesis that does not include equality as a possibility is referred to as the **alternative hypothesis**, written H_a.

The sample data give the only way that we have to distinguish between the null hypothesis and the alternative hypothesis. For example, if the null hypothesis claims that the population mean is 10 while the sample mean is 5 and the sample data are not very scattered, then the odds are good that the null hypothesis is wrong. Likewise, if the claim is that the population mean is greater than or equal to 10 and the sample mean is 5, then the odds are good that the null hypothesis is wrong, and similarly for the third case listed above.

In short, the sample data will be examined to see if there is a case for the alternative hypothesis to be true. In many ways the sample data act as evidence in a criminal trial, where the alternative hypothesis functions as a guilty verdict. If the evidence is strong enough (i.e., the sample data are extreme enough), then we will have to reject the null hypothesis in favor of the alternative hypothesis.

On the other hand, the sample data will never prove that the null hypothesis is true. It would be very difficult to convince someone that the population mean is exactly 10, no matter how much sample data we gather. The population mean might be 9.999 or 10.001 and we would still see the same kind of sample data.

Thus, if we do not reject the null hypothesis, we accept it rather than affirm it. Again, using the analogy of a criminal trial, a verdict of not guilty is not the same as a verdict of innocent—it merely indicates that there was not sufficient evidence to convict.

How do we measure when the sample data are extreme enough to contradict the null hypothesis? We need to use a test statistic, i.e., a random variable with a well-known probability distribution that can be used to measure how likely we are to get such sample data given that the null hypothesis is true. If, given the null hypothesis, the probability of getting such a sample is extremely low, then we would be inclined to reject the null hypothesis in favor of the alternative hypothesis. On the other hand, if the probability isn't too small, then the null hypothesis might well be true and we would have to accept it.

Common test statistics are the standard normal (Z) random variable, the **T random variable**, the **chi-square random variable**, (χ^2) and the **F random variable**.

Before we go ahead and evaluate the value of our test statistic (which is usually done with a statistical software package or a calculator), we need to set a significance level. The significance level is usually written as the letter α. This is our cutoff in terms of what we consider to be a probability that could happen by chance, and typically is either 5% or 1%.

For example, suppose that we set our significance level at 5%, evaluate our test statistic using the given sample data, and find out the probability associated with the value of our test statistic given that the null hypothesis is true. If that probability is less than 5%, then we consider it unlikely that we would get such sample data if the null hypothesis were true, and so we reject it. On the other hand, if the probability of getting such sample data were more than 5%, then we would accept the null hypothesis as reasonable.

The 5% functions as our level of reasonable doubt. If we used 1% instead of 5%, then it would be harder to reject the null hypothesis—the evidence would have to be stronger before we would do so.

It is important to choose the significance level before evaluating the test statistic—otherwise you might be tempted to use a significance level that would have you choose the hypothesis that you prefer to be true and the test would cease to be objective.

The probability that given the null hypothesis you would get sample data this extreme or worse is called the **p-value** of the test. Once the p-value is found, it is compared to the significance level. If the p-value is larger than the significance level, then you accept the null hypothesis. If the p-value is smaller than the significance level, then you reject the null hypothesis.

In order to evaluate the p-value, most software packages will need to know which of the three cases listed on page 170 that you are in. Specifically, the first case has two ways that the null hypothesis can be proved wrong, namely that the parameter is bigger than or smaller than the constant a. This kind of test is called a **two-tailed hypothesis test** because in the graph of the probability distribution of the test statistic, the "tails" to both the left and right correspond to rejection of the null hypothesis.

The second case, where rejection occurs because we think that the population parameter is smaller than the constant a, is called a **left-tailed hypothesis test**. Similarly, the third case, where rejection occurs because we think that the population parameter is larger than the constant a, is called a **right-tailed hypothesis test**. Collectively, the left-tailed and right-tailed hypothesis tests are called one-tailed tests.

There are two ways that we can come to the wrong conclusion even if all of the arithmetic has been done perfectly. We can reject the null hypothesis when it is correct. The chance that we will do so is exactly our significance level α. This kind of error has traditionally been known as a **Type I error**. The other way that we can go wrong is to accept the null hypothesis when it is wrong. This type of error is known as a **Type II error**, and the probability of it occurring depends on the true value of the population parameter. Since this can take many different values, this probability (often written as β) is a function of the true population parameter. See **power of test**.

Independence

- Two events A and B are said to be independent if knowledge of whether A has occurred does not change the likelihood of B occurring. In that case the **conditional probability** that A occurs, given that B has occurred, is the same as the ordinary probability of A: $\Pr(A \mid B) = \Pr(A)$.

- Two random variables X and Y are said to be independent if knowledge of the value of X does not change the probability distribution of Y. In that case the **covariance** and the **correlation coefficient** between the two random variables will both be zero.

Independent Events

See **independence**.

Independent Variable

In a **regression** model, the input variables are called independent variables. In simple regression, there is only one independent variable, for example, X in the equation

$$Y = aX + b$$

In multiple regression there is more than one independent variable, for example, X_1, X_2, and X_3 in the equation

$$Y = B_0 + B_1 X_1 + B_2 X_2 + B_3 X_3$$

Inferential Statistics

Refers to the use of sample data to make a claim about the population. Examples include **hypothesis testing**, where hypotheses about population parameters are tested, and **confidence intervals**, which are used to estimate population parameters.

Integral

An integral, specifically a definite integral, is a way of summing items over a continuous interval that is analogous to the sum $\sum_{k=1}^{n} f(k) = f(1) + f(2) + \ldots + f(n)$. This sum can be thought of as the area under rectangles of width one that have height $f(k)$. If we want to find the area under the curve $y = f(x)$ for $a \leq x \leq b$, then we take the integral of $f(x)$ from a to b, i.e.,

$$\int_a^b f(x)dx$$

This can often be calculated by finding an antiderivative $F(x)$ for $f(x)$, i.e., a function $F(x)$ such that $F'(x) = f(x)$. Then we have by the Fundamental Theorem of Calculus:

$$\int_a^b f(x)dx = F(b) - F(a)$$

(See Figure 9.13.)

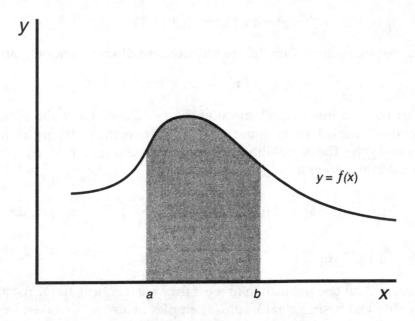

Figure 9.13: Definite integral $= \int_a^b f(x)dx =$ area

If X is a continuous random variable whose density function is $f(x)$, then the probability that X will be between two values a and b is given by the integral

$$\Pr(a < X < b) = \int_a^b f(x)dx$$

Unfortunately, it is usually not possible to find formulas for the antiderivatives needed to calculate the areas under the curves used in statistics, so it is necessary to consult tables or use computers.

Interpolation

Interpolation is the use of a **regression** model to make predictions with values of the independent variables that lie within the sample data. Interpolation is much more reliable than **extrapolation**.

Interquartile Range

The interquartile range is a measure of how spread out a collection of numerical data is. It is calculated as the difference between the third **quartile** and the first quartile.

Joint Probability Distribution

When several random variables $X_1, X_2, \ldots, X_n$ are defined, then probabilities for various combinations of their values can be calculated. Collectively, these probabilities form a joint probability distribution.

For two discrete random variables X and Y, the joint probability distribution is:

$$f(x,y) = \Pr[(X = x) \text{ and } (Y = y)]$$

For examples, see page 108. If X and Y are independent discrete random variables:

$$f(x,y) = \Pr(X = x) \times \Pr(Y = y)$$

If X and Y are two continuous random variables, then the joint probability function defines a hill whose total volume is 1; the volume under the hill over the rectangle from $x = a$ to $x = b$ and $y = c$ to $y = d$ gives the probability $\Pr[(a < X < b) \text{ and } (c < Y < d)]$. This volume can be defined by a double **integral**:

$$\Pr[(a < X < b) \text{ and } (c < Y < d)] = \int_{x=a}^{x=b} \int_{y=c}^{y=d} f(x,y) \, dy \, dx$$

Kruskal-Wallis H Test

When testing more than one population to see if they are identical, the Kruskal-Wallis H test is a **nonparametric** test based on rank sums. Samples of size n_i are taken from each of the k populations, and all of the measurements ranked and the ranks summed for each sample. Then

$$H = \frac{12}{n(n+1)} \sum_{i=1}^{k} \frac{R_i^2}{n_i} - 3(n+1)$$

where R_i is the rank sum for sample #i, and n is the sum of all of the n_i terms. If all of the n_i terms are over 5, then H is a **chi-square random variable** with $k-1$ **degrees of freedom** if the null hypothesis is true. The test is a right-tailed test.

Kurtosis

Kurtosis by definition is the expected value

$$\frac{E[(X - \mu)^4]}{\sigma^4}$$

Here μ is the mean and σ the standard deviation of the random variable X. Kurtosis provides a measure of how far out the tails of the distribution go.

Law of Large Numbers

The law of large numbers says that the sample mean is a consistent estimator for the population mean, i.e., that the variance of the sample mean approaches zero as the sample size increases.

Specifically, if $X_1, X_2, \ldots X_n$ are n independent random variables with identical probability distributions with mean μ and variance σ^2, then their average $\bar{x}$ will be a random variable with $E(\bar{x}) = \mu$ and $\text{Var}(\bar{x}) = \sigma^2/n$. Note how the variance of $\bar{x}$ becomes smaller as n becomes larger.

For example, the law of large numbers says that if you toss many dice (so that n approaches infinity), the average number that appears will approach 3.5 (which is μ, the expected value for one die). However, suppose you have already tossed 100 dice and found an average of

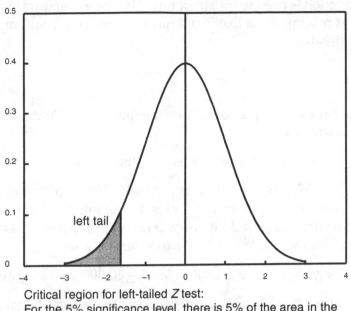

Critical region for left-tailed Z test:
For the 5% significance level, there is 5% of the area in the
left tail. The critical value is −1.65.

Figure 9.14: Critical region: left-tailed Z test

3.51. This does not mean that future tosses are more likely to be slightly less than 3.5 to balance it out. Each toss is still independent. Rather, the law of large numbers asserts that as you do many tosses, the fact that the average of the first 100 tosses differs slightly from 3.5 becomes unimportant. For example, if you toss the dice one million times, it should be clear that the first 100 tosses will not have much effect on the resulting average.

See also **Central Limit Theorem**.

Least Squares

The method of least squares chooses a **regression** model that will minimize the sum of the squares of the residuals.

Left-Tailed Hypothesis Test

A hypothesis test where the alternative hypothesis claims that a given population parameter is strictly less than a given value. In the graph of the probability distribution, the critical region for a left-tailed hypothesis test is on the left side (see Figure 9.14). For contrast, see **right-tailed hypothesis test**.

Likelihood Function

For a discrete random variable X, the likelihood function $L(x_1, x_2, \ldots, x_n)$ where $x_1, x_2, \ldots, x_n$ are values of sample data, is the product of the probabilities of X taking each of the values $x_1, x_2, \ldots, x_n$. If X is a continuous random variable, then $L(x_1, x_2, \ldots, x_n)$ is the product of the values of X's density function at each of those values.

The likelihood function is used to construct formulas for **estimators**, by finding values for the parameter(s) that maximize the likelihood function for the given sample data. See **maximum likelihood estimator**.

Logarithm

Let a be a positive number not equal to 1. Then the logarithm to base a is the inverse of the exponential function with base a:

$$\text{If } y = a^x, \text{ then } x = \log_a y$$

For example, since $2^5 = 32$, it follows that $\log_2 32 = 5$. The logarithm of the base itself is 1: $\log_a a = 1$, since $a^1 = a$. If $a = 10$ then $\log_{10}(x)$ is the common logarithm used for physical measurements such as pH and sound intensity. If no base is specified, as in $\log x$, then base 10 is usually assumed. Some examples of base 10 logarithms are:

$$\log 10 = 1; \quad \log 100 = 2; \quad \log 1{,}000 = 3; \quad \log 10{,}000 = 4; \quad \log 100{,}000 = 5$$

In mathematics the base $e \approx 2.718281828459045$ occurs frequently. Logarithms to base e are called natural logarithms and are often symbolized by $\ln x$.

Logarithms to any base satisfy these properties:

$$\log(xy) = \log x + \log y$$
$$\log(x/y) = \log x - \log y$$
$$\log(x^n) = n \log x$$
$$\log 1 = 0$$
$$\log x > 0 \text{ for } x > 1 \text{ if the base is greater than 1}$$
$$\log x < 0 \text{ for } 0 < x < 1 \text{ if the base is greater than 1}$$

These properties are very helpful in regression analysis, where curved relationships can be transformed into straight line relations. If the true relation between x and y is of the form:

$$y = ab^x$$

then take the logarithm of both sides:

$$\log y = \log a + x \log b$$

This becomes a linear regression, with $\log a$ as the constant term in the regression, $\log b$ the slope, and $\log y$ used (instead of y itself) as the dependent variable. If the relation is of this form:

$$y = mx^n$$

transform it as follows:

$$\log y = \log m + n \log x$$

This becomes a linear regression with $\log m$ as the constant, n as the slope, $\log y$ as the dependent variable, and $\log x$ (instead of x itself) as the independent variable.

The same idea can be extended to multiple regression. The relation

$$y = B_0 X_1^{B_1} X_2^{B_2} X_3^{B_3}$$

can be transformed into linear form:

$$\log y = \log B_0 + B_1 \log X_1 + B_2 \log X_2 + B_3 \log X_3$$

Mann-Whitney U Test

Given independent samples from two populations, it is possible to perform a **nonparametric** hypothesis test as to whether the populations are identical. The observations are ranked, and the sum of the ranks for each of the two populations is calculated. Let W_1 be the rank sum for the first population, and let n_1 and n_2 be the two sample sizes. Then the Mann-Whitney U statistic is given by:

$$U = n_1 n_2 + \frac{n_1(n_2 + 1)}{2} - W_1$$

For small sample sizes, critical values for U can be found from tables. If both sample sizes are over 10, then we can construct

$$Z = \frac{U - n_1 n_2 / 2}{\sqrt{n_1 n_2 (n_1 + n_2 + 1)/12}}$$

Marginal Probability Function

Given a **joint probability distribution** $f(x, y)$ for two random variables X and Y, the marginal probability function for $X(f_X(x))$ is obtained by summing over all of the possibilities for Y if it is discrete, or by integrating over all of the possibilities for Y if it is continuous.

$$\text{discrete:} \quad f_X(x) = \sum_{i=1}^{m} f(x, y_i)$$

The values $y_1, y_2, \ldots y_m$ are the m possible values of Y.

$$\text{continuous:} \quad f_X(x) = \int_{y=-\infty}^{y=\infty} f(x, y)dy$$

The marginal probability of X is the same as the ordinary probability function one would observe for X if Y were not in the picture.

If X and Y are independent, then their two marginal probability functions can be multiplied to give the joint probability function:

$$f(x, y) = f_X(x) \times f_Y(y)$$

Markov Chain

A Markov Chain is a series of events drawn from a fixed list of possibilities, where the probabilities of going from one state to another are fixed. An example would be a six-sided die that is rolled from one side to another. There are six possible states. If there is a 1 on top, then there is a 1/4 chance that rolling the die over will place either a 2, 3, 4, or 5 on top, and no chance that 1 or 6 will appear on top at the next stage.

Maximum Likelihood Estimator

Once a **likelihood function** has been constructed, it is possible to solve for the value of a parameter that maximizes the function, given the sample data. The resulting formula is the maximum likelihood estimator for that parameter. For example, the sample average $\bar{x}$ is the maximum likelihood estimator for the population mean μ when the population is normal.

Mean

The mean of a collection of numbers (whether the data is from a population or a sample) is the average of those numbers, obtained by summing the numbers and dividing by n, where n is how many numbers there are in the collection. The mean of a population is typically represented by μ. The mean of a sample is typically represented by $\bar{x}$.

Median

The median of a collection of numbers (whether the data is from a population or a sample) is the halfway point in the following sense: half of the numbers lie below the median, and half of them lie above it.

To calculate the median, first put the numbers in order. Let n be how many numbers there are in the collection. If n is odd, then the number in the $\frac{(n+1)}{2}$ position is the median; if n is even, then the average of the two numbers in the $\frac{n}{2}$ and $\left(\frac{n}{2}\right)+1$ positions is the median.

The median is the second **quartile** and the fiftieth **percentile**.

If you have frequency distribution data for different classes (rather than the original numbers themselves), you can estimate the median from this formula:

$$\text{median} = x_L + \frac{(N/2 - n_L)w}{n_m}$$

Here x_L is the lower limit of the median class; N is the total number of items; n_L is the number of items below the median class; n_m is the number of items in the median class; and w is the width of the median class.

For example, here is a frequency table:

Class	Frequency	Cumulative Frequency	Cumulative Relative Frequency
0 − 1,000	30	30	.15
1,000 − 2,000	60	90	.45
2,000 − 3,000	70	160	.80
3,000 − 4,000	40	200	1.00

We first determine that 2,000 to 3,000 is the median class. The median class will always be the first class where the cumulative relative frequency is greater than .5. Then use the formula:

$$\text{median} = 2,000 + \frac{(200/2 - 90)1,000}{70} = 2,000 + \frac{10 \times 1,000}{70} = 2,142.9$$

Mode

A mode is a number that occurs most frequently in a collection of numbers. If there is a tie for which number occurs most frequently, then there is more than one mode. In a frequency histogram, the mode occurs at the highest bar.

Model

A model is a description of how a population behaves. As such, a model includes information on the probabilities associated with a population. Examples include a **normal random** variable, a **Poisson random variable**, a **geometric random variable**, and a **binomial random variable**.

Moment

The nth moment of a random variable X is the average of X^n, i.e., $E[X^n]$.

Mu

The Greek letter mu (μ) is used to stand for the mean of a population or of a random variable.

Multicollinearity

In constructing a **regression** model, if the independent variables are not truly independent of each other then multicollinearity exists. The coefficients become indeterminate when this occurs, and large numerical errors may occur. You can tell the problem will occur if two of the independent variables have a high correlation coefficient (near 1).

For example, suppose you are performing a multiple regression using observations of people where income and education are two of the independent variables. If all of the high education people in your sample have high income, and vice versa, then the problem of multicollinearity arises and you will be unable to distinguish the effect of income separately from the effect of education.

Multinomial Random Variable

A multinomial random variable is a generalization of a binomial random variable. Suppose that there are n independent trials, each with k fixed outcomes with known probabilities. Let X_i count the number of trials with outcome #i; then X_i is a multinomial random variable, and collectively the X_i's form a multinomial distribution. The **probability function** for a multinomial distribution is given by

$$\Pr(X_1 = n_1, X_2 = n_2, \ldots, X_k = n_k) = \frac{n!}{n_1!n_2!\ldots n_k!}p_1^{n_1}p_2^{n_2}\ldots p_k^{n_k}$$

The quantity p_i is the probability of X_i occurring, and $n = n_1 + n_2 + \ldots + n_k$.

For example: suppose that a fair six-sided die is rolled. Let X_1 be the event that a 1 is rolled, X_2 be the event that a 2 or a 3 is rolled, and X_3 be the event that a 4, 5, or 6 is rolled. Then together X_1, X_2, and X_3 form a multinomial distribution with **probabilities** $p_1 = 1/6$, $p_2 = 2/6$, and $p_3 = 3/6$. The probability that out of 10 rolls, X_1 occurs two times, X_2 occurs three times, and X_3 occurs five times is given by:

$$\Pr(X_1 = 2, X_2 = 3, X_3 = 5) = \frac{10!}{2!3!5!}\left(\frac{1}{6}\right)^2\left(\frac{2}{6}\right)^3\left(\frac{3}{6}\right)^5 = .081$$

Multiple Regression

Consider a situation where one dependent variable (Y) can be expressed as a linear function of several independent variables ($X_1, X_2, \ldots X_{m-1}$) of this form:

$$Y = \beta_0 + \beta_1 X_1 + \beta_2 X_2 + \ldots + \beta_{m-1} X_{m-1}$$

The true values of the β's are unknown; however, you do have n observations of each of the variables. Multiple regression analysis is used to estimate the values of the coefficients that provide the best fit to this equation. Note that there are m coefficients to estimate: one for each of the $m-1$ independent variables, plus the constant term β_0. (For the case where there is only one independent variable, see **simple regression**.) It is assumed that there is a random variable called the error term that accounts for variations in Y that are not explained by the equation. The error term has zero mean; it is often assumed to have a normal distribution with an unknown variance. That variance should be small if the multiple regression equation is to be very reliable.

The regression calculation reports a set of values $B_0, B_1, \ldots, B_{m-1}$ that are used as estimators for the true values $\beta_0, \beta_1, \ldots, \beta_{m-1}$. Once values for the coefficients have been found, it is possible to find an estimated value $\hat{Y}_i$ for each set of values of the independent variables:

$$\hat{Y}_i = B_0 + B_1 X_{i1} + B_2 X_{i2} + B_3 X_{i3} + \ldots + B_{m-1} X_{i,(m-1)}$$

The quantity X_{ij} is the ith observation of the variable X_j. This estimated value can be compared with the true value Y for that observation. The goal of multiple regression is to minimize the sum of the squared deviations between the true value Y_i and the estimated value $\hat{Y}_i$:

$$\text{squared error} = \sum_{i=1}^{n} (Y_i - \hat{Y}_i)^2$$

The calculation procedure for the regression coefficients is very complicated, requiring a knowledge of matrix algebra. If the independent variables are arranged into a matrix $\mathbf{X}$ of n rows and m columns (where each column represents the n observations of one of the independent variables, and one column consists solely of 1's, to take account of the constant term), and the dependent variable observations are arranged into an n by 1 matrix $\mathbf{Y}$, then the m-by-1 coefficient matrix $\mathbf{B}$ comes from this formula:

$$\mathbf{B} = (\mathbf{X}^{tr}\mathbf{X})^{-1}\mathbf{X}^{tr}\mathbf{Y}$$

The matrix $\mathbf{X}^{tr}$ is the transpose of $\mathbf{X}$, and $(\mathbf{X}^{tr}\mathbf{X})^{-1}$ is the inverse of the matrix formed by multiplying $\mathbf{X}^{tr}$ by $\mathbf{X}$.

Fortunately, in practice you do not ever need to work with the formula, because a computer will do the calculations for you. The computer will report estimated coefficient values, as well as an r^2 value that tells you whether there is a good fit for the data. As with simple regression, an r^2 value of 1 means a perfect fit; an r^2 value of 0 means that 0% of the variation in the dependent variable is accounted for by variations in the independent variables.

The computer will also report a standard error for each coefficient; a larger value for the standard error means that there is more uncertainty about the true value of that coefficient. Dividing the estimated coefficient by the corresponding standard error gives a quantity known as the t statistic, which is used for hypothesis tests about whether or not a particular variable belongs in the regression equation. If the true value of that coefficient is zero, then the t statistic will come from a t distribution with $n-m$ degrees of freedom (n is the number of observations; m is the number of coefficients that are estimated, including the constant term). If the absolute value of the reported t statistic is greater than the absolute value of the critical value from the t distribution table, then reject the null hypothesis—the true coefficient is not zero, and the variable belongs.

The computer will also report an F statistic, which is used to test the hypothesis that the coefficients of all of the independent variables are zero. If the null hypothesis is true, and the coefficients are all zero (meaning the regression calculation is worthless for predicting the value of Y), then the F statistic will come from an F distribution with $m - 1$ numerator degrees of freedom and $n - m$ denominator degrees of freedom. If the reported value is greater than the critical value for those degrees of freedom, then reject the null hypothesis.

In practice the difficult matter with regression analysis is determining which variables to include, and the exact form to use for the equation. You can test to see whether a variable that is included really belongs; however, you might have left out variables that should be included. The restriction that the equation be linear is not a big problem; if the true relationship involves a quadratic curve, such as $Y = aX_1 + bX_1^2$, then simply include both X_1 and X_1^2 as independent variables in your regression analysis. If the true relation is of the form $Y = X_1^{B_1} X_2^{B_2}$, see **logarithm** for information on transformations that convert it to a linear form.

Some other problems that can arise with multiple regression include **multicollinearity** (when two or more of the independent variables are highly correlated); **heteroscedasticity** (when the variances of the error terms are different for different observations); and **serial correlation** (when the errors for successive observations of time series data are correlated with each other). These problems make it more difficult for the regression calculation to estimate the coefficients accurately.

Here are some sample data:

X_1	X_2	X_3	X_4	Y
2	1	3	0	12
3	3	8	0	23
2	5	3	1	29
2	5	7	-1	27
5	1	3	-1	20
5	1	8	0	21
4	6	4	1	39
5	5	8	0	37

The value of Y is given by the equation:

$$Y = 2 + 3X_1 + 4X_2 + X_4$$

However, remember that in reality we will not be able to see the true equation as we can in this artificial example. If we knew that X_1, X_2, and X_4 all should be included, then we would run the regression with those independent variables and we would find an r^2 value of 1, with each of the true coefficients found exactly. However, in reality we don't know for sure which variables should be included. Suppose we perform a multiple regression calculation with X_1 and X_2 as the independent variables. The results are:

Variable	X_1	X_2	Constant term
Coefficient	2.9780	4.1538	1.5581
Standard error	0.2191	0.1452	1.0339
t statistic	13.5890	28.6001	1.5070

The r^2 value is 0.9943. We do not have a perfect fit, because we left out the variable X_4. However, X_4 has only a very small influence on Y, so leaving it out has not hurt our regression equation noticeably. The estimated coefficients (2.978 and 4.1538) are close to the true values (3 and 4, respectively). The t statistics need to be compared against a t distribution with $8 - 3 = 5$ degrees of freedom, which gives a critical value of 2.571 using the 5% significance level (page 209). The two t statistics (13.589 and 28.6) are way above the critical value, so we can clearly reject the hypothesis that the true coefficients are zero.

The F statistic for this regression is reported to be 438.4; this needs to be compared against an F distribution with $3 - 1 = 2$ numerator degrees of freedom and $8 - 3 = 5$ denominator degrees of freedom. The critical value is 5.79 (page 211). The observed value is way above this limit, so the null hypothesis that both coefficients are truly zero can clearly be rejected.

Now suppose we perform a regression calculation with X_2 and X_3 as the independent variables. We know from the true equation that X_3 doesn't belong in the equation, but X_1 and X_4 do. Unfortunately, the researcher in the field does not see the true equation, and will not always know if important variables have been left out. In this case the resulting regression equation is:

Variable	X_2	X_3	Constant term
Coefficient	3.6834	0.4550	11.0663
Standard error	0.8464	0.7373	5.0216
t statistic	4.3520	0.617	2.2038

The r^2 value falls to .8001. The estimated coefficient for X_2 is still close to its true value of 4; its t statistic (4.352) is still above the critical value so we reject the null hypothesis that the true coefficient of X_2 is zero. The t statistic for X_3 falls inside the interval -2.571 to 2.571, so we accept the null hypothesis that the true coefficient of X_3 is zero. This happens to be correct, because we know that X_3 is not included in the true equation. However, the regression results provide no way of testing for the fact that there are variables that should be included (X_1 and X_4) but are missing. In this case the missing variables do not hurt us too badly, but in other cases missing variables can wreak havoc on our ability to estimate the coefficients of the variables that are included.

Mutually Exclusive

Mutually exclusive is a synonym for **disjoint**. Two events A and B are mutually exclusive if they cannot both happen, so $\Pr(A \text{ and } B) = 0$.

Negative Binomial Random Variable

An experiment with two possible outcomes, success (with probability p) and failure (with probability $q = 1 - p$) is repeated independently until there are k successes. The number of trials X that it takes is a negative binomial random variable with parameters k and p. Since the last trial must be a success for the experiment to stop, the other $k - 1$ successes must be arranged among the $n - 1$ trials preceding the last one, where $X = n$. Then

$$\Pr(X = n) = \binom{n - 1}{k - 1} p^k q^{n-k}$$

The negative binomial random variable is discrete, and can take any integer value from k onward. Its mean is k/p and its variance is kq/p^2. See also **geometric random variable** and **binomial random variable**.

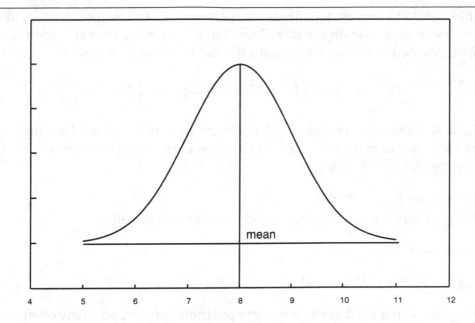

4 5 6 7 8 9 10 11 12

Figure 9.15: Probability density function for normal random variable

Nonparametric

Nonparametric statistics refers to the analysis that can be made on sample data without assuming foreknowledge of the population's distribution. Examples include the **sign test**, the **Wilcoxon signed rank test**, the **Mann-Whitney U test**, and the **Kruskal-Wallis H test**.

Normal Random Variable

The normal random variable is one of the most important and frequently occurring models because of the **Central Limit Theorem**. The graph of its density function is bell-shaped (see Figure 9.15).

The normal random variable is determined by two parameters: its mean (denoted μ) and its variance (denoted σ^2). The normal random variable can take any value, although values farther from the mean become less likely. Its probability density function is given by

$$f(x) = \frac{1}{\sigma\sqrt{2\pi}}e^{\frac{-(x-\mu)^2}{2\sigma^2}}$$

The peak of the density function curve occurs at μ. Here are two properties of normal random variables:

- If X and Y are independent normal random variables, then $X + Y$ also will be a normal random variable (with mean $\mu_x + \mu_y$ and variance $\sigma_x{}^2 + \sigma_y{}^2$).

- If X is a normal random variable and a and b are constants, then $aX + b$ will also be a normal random variable (with mean $a\mu_x + b$ and variance $a^2\sigma_x{}^2$). In particular, $\frac{(X-\mu)}{\sigma}$ will be a normal random variable with mean 0 and standard deviation 1.

If the mean is 0 and the standard deviation is 1, then the variable is called a standard normal random variable, or Z random variable. Probabilities involving normal random variables can always be expressed in terms of a standard normal (Z) random variable:

$$\Pr(X < a) = \Pr\left(\frac{X - \mu}{\sigma} < \frac{a - \mu}{\sigma}\right) = \Pr\left(Z < \frac{a - \mu}{\sigma}\right)$$

The variable X is a normal random variable with mean μ and standard deviation σ. The probability that Z is less than a particular value is given in the standard normal table (page 205).

Some properties of Z include:

- $\Pr(Z > a) = 1 - \Pr(Z < a)$
  (This property is shared by any continuous random variable.)

- $\Pr(Z > a) = \Pr(Z < -a)$

- $\Pr(Z < -a) = 1 - \Pr(Z < a)$

It is important to realize that while many populations are normal, many others are not. You should examine the frequency histogram for a sample before applying any hypothesis tests or calculating any confidence intervals that assume normality. If the frequency histogram is not bell-shaped, then you will need to use techniques that do not require normality.

Null Hypothesis

The null hypothesis, usually denoted H_0, is one of two complementary hypotheses present in a hypothesis test. The null hypothesis is the one of the two that contains equality as a possibility. For example, each of the following is a possible null hypothesis:

- The population mean is equal to 3.

- The population mean is less than or equal to 3.

- The population mean is greater than or equal to 3.

For contrast, see **alternative hypothesis**.

Numerator Degrees of Freedom

The **F random variable** has two different degrees of freedom: one for the numerator and the other for the denominator. This occurs because formally the F random variable is defined to be proportional to the ratio of two **chi-square random variables**, each with its own degrees of freedom. In practice, the context will determine both the numerator degrees of freedom and the denominator degrees of freedom for an F random variable.

Ogive

An ogive consists of points representing cumulative frequencies for classes connected by line segments (see Figure 9.16).

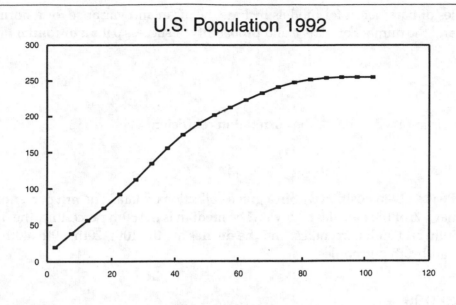

U.S. Population 1992

Figure 9.16: Ogive

One-Tailed Hypothesis Test

A one-tailed hypothesis test is either a left-tailed hypothesis test or a right-tailed hypothesis test. In either case, there will be only one critical region and only one critical value.

Outcome

An outcome is one possible result from an experiment. See also **event**.

P-Value

When a **hypothesis test** is performed, the p-value is the probability of getting a sample as extreme as the given one or worse, given that the null hypothesis is true. If the p-value is very small (specifically, if it is smaller than the significance level), then the null hypothesis should be rejected.

For example, suppose you are performing a right-tailed hypothesis test based on a standard normal (Z) random variable. If the test statistic value is $Z = 3$, then the p-value is .0013 (the area to the right of 3, found from the table on page 205). A p-value this small gives you strong evidence to reject the null hypothesis.

If you were performing a two-tailed test and found a test statistic of 3, then the p-value would be .0026 (the area in both tails that are outside the range -3 to 3).

Parameter

A parameter is a number that is associated with a population or a model of a population. Usually it is impossible to find the exact value of a population parameter; instead an **estimator** based on sample data is used.

Examples of parameters for models include the mean and variance for a **normal random variable**, and the number of trials n and probability of success p for a **binomial random variable**.

Pearson's r

Pearson's r is another name for the **correlation coefficient**.

Percentile

A percentile is a given position within a given collection of data. The nth percentile is a number such that $n\%$ of the data lies below it. The **median** is the 50th percentile, the **quartiles** are the 25th, 50th, and 75th percentiles, and the **deciles** are the 10th, 20th, 30th, 40th, 50th, 60th, 70th, 80th, and 90th percentiles.

Permutations

The permutations of k objects chosen from a group of n objects refers to the number of ways of choosing the objects where the order of selection matters. The number of permutations is given by

$$P_k^n = \frac{n!}{(n-k)!}$$

See also **combinations**.

Pie Chart

A pie chart is a circular display of proportions; the relative size of a category corresponds to the size of its pie slice (see Figure 9.17).

Poisson Random Variable

The Poisson random variable is a limit of the **binomial random variable** as the number of trials goes off to infinity. It takes values of $0, 1, 2, \ldots$, and is appropriate for modeling the count of successes when the probability of success is small but the number of trials is large. The Poisson random variable is determined by one parameter, λ, and the graph of its probability distribution is skewed to the right. Its probability function is given by

$$\Pr(X = k) = \begin{cases} \frac{\lambda^k e^{-\lambda}}{k!} & k \geq 0 \\ 0 & \text{otherwise} \end{cases}$$

Its mean and variance are both equal to λ (see Figure 9.18).

Population

A population is a collection of all data points of interest. Usually the population is too large to be worked with directly; rather, a sample is taken and analyzed.

Great Lakes: area (square miles)	
Superior	31,700
Huron	23,000
Michigan	22,300
Erie	9,910
Ontario	7,340

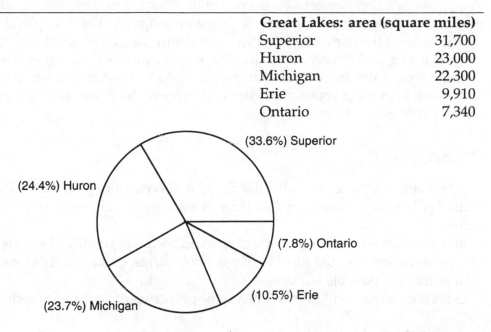

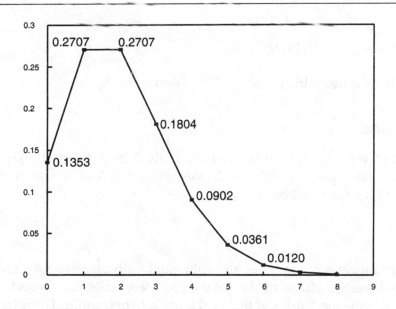

Figure 9.17: Pie chart

Figure 9.18: Poisson distribution with lambda = 2

Power of Test

The power of a hypothesis test is the probability that the null hypothesis will be rejected, expressed as a function of the parameter being investigated. For example, if you were testing for the value of the population mean μ, then ideally the power function would equal 0 at the true value of μ, and 1 everywhere else. This power function would guarantee that the correct decision would always be made, but you can seldom expect such a nice situation in practice. In general, increasing your sample size will improve the power function by making it more like the ideal power function.

Probability

The probability of an event is the likelihood of it occurring. This likelihood is expressed as a number between 0 (certainty that the event will not occur) and 1 (certainty that the event will occur).

In a situation where all outcomes of an experiment are equally likely, then the probability of an event is equal to the number of outcomes corresponding to that event divided by the total number of possible outcomes.

Let A and B represent events. Then some properties of probability include:

- $\Pr(not A) = 1 - \Pr(A)$

- $\Pr(A \text{ or } B) = \Pr(A) + \Pr(B)$
  if A and B are mutually exclusive events.

- $\Pr(A \text{ or } B) = \Pr(A) + \Pr(B) - \Pr(A \text{ and } B)$
  for any two events.

- $\Pr(A \text{ and } B) = \Pr(A) \times \Pr(B)$
  if A and B are independent events.

- $\Pr(A \text{ and } B) = \Pr(A \mid B) \Pr(B)$

See also **conditional probability**.

Probability Function

The probability function for a **discrete random variable** X is a function $f(x)$ such that $f(x_i) = \Pr(X = x_i)$. If x_i is not a possible value of X, then $f(x_i) = 0$. The sum of all nonzero values of the probability function must be 1.

Quartile

If a collection of numbers is put in order, and split into fourths, the quartiles are the boundary points. There are three quartiles: the first is the 25th **percentile**, the second is the 50th percentile (which is also the median), and the third is the 75th percentile. The difference between the third and first quartile is called the **inter-quartile range**.

R Squared

The efficacy of a **regression** model is often measured by how much of the variability of the data that it explains. The ratio of the sums of squared deviation of the model to the actual sums of squared deviations in the data is denoted by r^2, and is also known as the coefficient of determination.

The value of r^2 by definition runs between 0 and 1; the closer to 1, the better the job the model does in explaining the given data.

Since r^2 can often be increased by adding more variables to the regression model, often an **adjusted r squared** is used that takes into account how many variables have been included.

Random Sample

In order to be useful in most analyses, sample data must be collected in such a way that all population data stand the same chance of being included in the sample. Such a sample is called a random sample.

Random Variable

A random variable is any measurement that can be taken, numerical or otherwise. Prior to the measurement being taken, the value of the random variable is unknown, but the probabilities of taking certain values are often known. See **discrete random variable** and **continuous random variable**.

Range

The range is a direct measure of the spread of a collection of numbers: it is the difference between the largest and smallest numbers in the collection.

Regression

Regression is a process by which an equation is found to describe the relationship between several random variables on the basis of sample data. The inputs to the equation are the independent variables, while the output from the equation is the dependent variable.

If there is only one independent variable, then the process is known as **simple regression**; if there is more than one independent variable, then the process is known as **multiple regression**.

Relative Frequency

For a given class, the relative frequency is the fraction of data points that are contained in that class. The relative frequency is found by dividing the frequency for the given class by the number of data points, and is usually expressed as a percentage.

Relative Frequency Histogram

A relative frequency histogram is a bar chart (usually drawn vertically) indicating the relative frequencies for the various classes of data. See Figure 9.19 for an example.

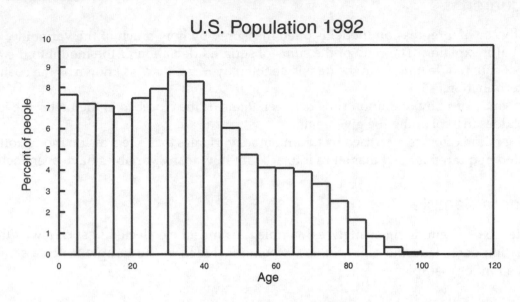

Figure 9.19: Relative frequency histogram

Residual

The residual of a point in **regression** analysis is the difference between the true value of the dependent variable and its predicted value.

Right-Tailed Hypothesis Test

A right-tailed hypothesis test is a **hypothesis test** where the alternative hypothesis claims that a given population parameter is strictly greater than a given value. In the graph of the probability distribution, the critical region for a right-tailed hypothesis test is on the right side (see Figure 9.20). For contrast, see **left-tailed hypothesis test**.

Sample

A sample is a small collection of data, taken from a population, that is more amenable to analysis. A good sample should be representative of its population; this can be achieved if the sample is a random sample.

Sample Data

The sample data are the information contained within a sample.

Sample Points

The sample points are the components of the sample data that are linked together. For example, in a sampling of adult human weights and heights, a sample point would be the weight and height of one person.

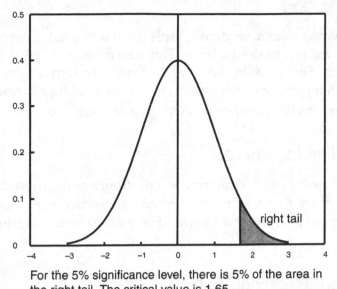

For the 5% significance level, there is 5% of the area in the right tail. The critical value is 1.65.

Figure 9.20: Critical region for right-tailed Z test

Sample Size

The sample size refers to the number of sample points. The greater the sample size, the more information that the sample contains. Larger samples allow you to have narrower **confidence intervals** and increase the power of your hypothesis tests.

Sample Statistic

A sample statistic is any number that is calculated from a sample. An example is the sample mean $\bar{x}$. A sample statistic is often used as an **estimator** for an unknown population **parameter**.

Sample Variance

The **variance** of the data in a sample, calculated from this formula:

$$s^2 = \frac{(X_1 - \bar{x})^2 + (X_2 - \bar{x})^2 + \ldots + (X_n - \bar{x})^2}{n - 1}$$

Sampling

Sampling is the process of collecting a sample, and is one of the most important components of statistics. Samples must in some way be representative of their populations, and random sampling is usually the best way to achieve this. See also **sampling with replacement** and **sampling without replacement**.

Sampling Distribution

If a statistic is calculated from a random sample, then it is itself a random variable whose probability distribution can be determined. This distribution is called the sampling distribution of that statistic. For example, if $X_1, \ldots X_n$ form a random sample chosen from a normal distribution with mean μ and variance σ^2, then the sampling distribution of the average $\overline{x} = \frac{(X_1 + \ldots + X_n)}{n}$ is a normal distribution with mean μ and variance σ^2/n.

Sampling With Replacement

Sampling with replacement refers to taking a sample from a population in such a way that the same object can be chosen twice. While uncommon in practice, sampling with replacement will usually simplify the analysis of a sample. For contrast, see **sampling without replacement**.

Sampling Without Replacement

Sampling without replacement refers to taking a sample from a population in such a way that the same object cannot be chosen twice. The number of **combinations** $\binom{n}{j}$ gives the number of ways of selecting j objects without replacement from a group of n objects. See also **hypergeometric distribution**.

Scatterplot

Given two numerical random variables X and Y based on the same sample points, the scatterplot is the plot of those sample points with X corresponding to the horizontal axis, and Y corresponding to the vertical axis. Unit 7 includes several examples of scatterplots. A scatterplot is also sometimes called an XY diagram.

Serial Correlation

Serial correlation refers to the correlations between regularly timed events. If a regression analysis uses time series data where the error in one time period is correlated with the error in the next period, then the problem of serial correlation arises, which adds to the uncertainty in the estimates of the true regression coefficients. The **Durbin-Watson** statistic is one way to test for this.

Sigma

The lower case Greek letter sigma (σ) is used to stand for the **standard deviation** of either a population or a random variable. Sigma squared (σ^2) is used to represent the **variance**.

The upper case Greek letter sigma (Σ) is used to represent summation. See **summation notation**.

Sign Test

When sample points consist of pairs of observations, it is possible to perform a **nonparametric** hypothesis test on the hypothesis that one observation is more likely to be the larger of the two, merely by noting which of the observations is larger for each sample point and comparing the result with that of a **binomial random variable** with $p = 1/2$.

Significance Level

In a **hypothesis test**, a cutoff needs to be set at the beginning as to what probabilities are likely and what probabilities are unlikely. The significance level is that cutoff, which represents the probability of a **type I error**.

In performing a hypothesis test, a **p-value** is found that is the probability of getting such an extreme sample or worse if the null hypothesis is true. If the p-value is less than the significance level, then it is unlikely that the null hypothesis is true, and we reject it. If the p-value is larger than the significance level, then the sample is not too unlikely, and we accept the null hypothesis.

Significance levels are usually set at 5% or 1%. The smaller the significance level, the harder it is to reject a null hypothesis.

Simple Regression

Simple regression involves finding the equation of a line that best fits a pattern of observations of two variables: an independent variable X and a dependent variable Y. Assume that the two variables are related by a linear equation of this form:

$$Y = a + bX$$

The slope of the line is b, and a is the y intercept, or constant term. However, the values of a and b are unknown. If you have n observations each for X and Y, you can determine the values of a and b that give the line that best fits the observations.

Let $\hat{a}$ and $\hat{b}$ be the estimated values for a and b that result from your regression calculation. Then, for each value X_i there is a corresponding predicted value of Y (call it $\hat{Y}_i$):

$$\hat{Y}_i = \hat{a} + \hat{b}X_i$$

The values of $\hat{Y}$ all lie along the regression line. In each case we can determine the difference between the predicted value of Y and the actual value of Y associated with that value of X (call this the error):

$$\text{error}_i = Y_i - \hat{Y}_i$$

By squaring each of these errors and adding, we can find SE_{line}, the total squared error about the regression line:

$$SE_{line} = \sum_{i=1}^{n}(Y_i - \hat{Y}_i)^2$$

The goal of regression analysis is to choose the values of $\hat{a}$ and $\hat{b}$ that minimize the value of SE_{line}. These values are called the ordinary least squares estimators of a and b, and they are found from these formulas:

$$\text{slope} = \hat{b} = \frac{\overline{xy} - \overline{x}\ \overline{y}}{\overline{x^2} - \overline{x}^2}$$

$$\text{constant term} = \hat{a} = \bar{y} - \hat{b}\bar{x}$$

The bar over each quantity designates that it is an average. In most practical situations, you will use a computer to perform the calculations.

Here are some sample data:

X	Y
7	6
5	4
3	9
2	8
10	1

In reality you would not want to perform these calculations with only five observations; more observations would give you better estimators.

To find the slope and intercept of the regression line, set up a table like this:

	X	Y	XY	X^2	Y^2
	7	6	42	49	36
	5	4	20	25	16
	3	9	27	9	81
	2	8	16	4	64
	10	1	10	100	1
Average:	5.4	5.6	23	37.4	39.6

Now use the formulas:

$$\text{slope} = \hat{b} = \frac{23 - 5.4 \times 5.6}{37.4 - 5.4^2} = -0.87864$$

$$\text{constant term} = \hat{a} = 5.6 - (-0.87864 \times 5.4) = 10.3447$$

Figure 9.21 shows the scatterplot for our five observations, with the regression line drawn in.

We also need to be able to measure whether this line fits the data very well. To do this, we calculate the r^2 value:

$$r^2 = 1 - \frac{SE_{line}}{SE_{avg}}$$

The quantity SE_{avg} is the squared error of the Y values about their average. The formula has the result that r^2 is always between 0 and 1, or $0 \leq r^2 \leq 1$. We can calculate r^2 from the following table:

X	Y	$\hat{Y}$	error = $Y - \hat{Y}$	$(Y - \hat{Y})^2$	$Y - \bar{Y}$	$(Y - \bar{Y})^2$
7	6	4.194	1.806	3.261	0.4	0.16
5	4	5.951	−1.95	3.808	−1.6	2.56
3	9	7.709	1.291	1.667	3.4	11.56
2	8	8.587	−0.59	0.345	2.4	5.76
10	1	1.558	−0.56	0.312	−4.6	21.16
			$SE_{line} =$	9.393	$SE_{avg} =$	41.2

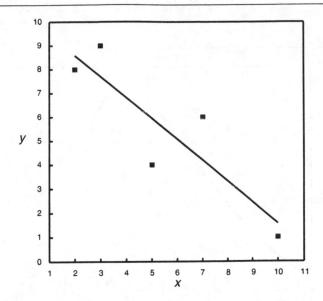

Figure 9.21: Scatterplot with regression line

We calculate $\hat{Y}$ from the equation $10.3447 - .87864X$. The SE_{line} and SE_{avg} values are the sums of their respective columns. Now we can calculate r^2:

$$r^2 = 1 - \frac{9.393}{41.2} = .772$$

This value means that 77.2% of the variation in Y is accounted for by variations in X. An r^2 value of 1 would mean a perfect fit; an r^2 value of zero would mean that you could predict Y just as well without knowing X as you can by knowing the regression equation.

The value of r^2 can also be calculated from this formula, since it is the square of the **correlation coefficient**:

$$r^2 = \frac{(\overline{xy} - \overline{x}\ \overline{y})^2}{(\overline{x^2} - \overline{x}^2)(\overline{y^2} - \overline{y}^2)} = \frac{(23 - 5.4 \times 5.6)^2}{(37.4 - 5.4^2)(39.6 - 5.6^2)} = .772$$

For situations where there is more than one independent variable, see **multiple regression**. If the true relation between x and y is of the form $y = ab^x$ or $y = mx^n$, see **logarithm**.

Skewed to the Left

A graph of a probability distribution or a histogram is skewed to the left if it trails off slowly to the left, with the bulk of the graph or histogram lying to the right. In either case, the mean will lie to the left of the median (see Figure 9.22).

Skewed to the Right

A graph of a probability distribution or a histogram is skewed to the right if it trails off slowly to the right, with the bulk of the graph or histogram lying to the left. In either case, the mean will lie to the right of the median (see Figure 9.23).

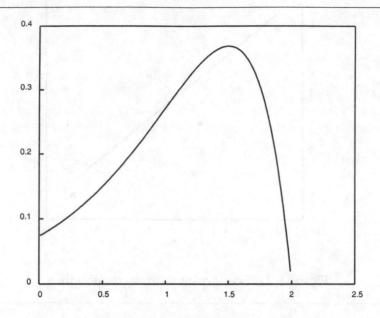

Figure 9.22: Skewed to the left

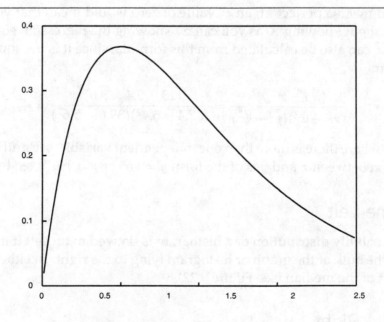

Figure 9.23: Skewed to the right

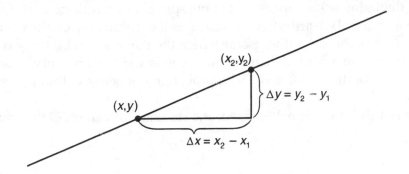

$$\text{Slope of line} = \frac{\Delta y}{\Delta x}$$
$$\Delta = \text{delta} = \text{``change in''}$$

Figure 9.24

Skewness

Formally, the skewness of a random variable X with mean μ and standard deviation σ is given by:

$$\frac{E[(X - \mu)^3]}{\sigma^3}$$

If the skewness is positive, then the distribution is **skewed to the right**. If it is negative, then the distribution is **skewed to the left**.

Slope

The slope of a line is a measure of the line's direction. It can be calculated by choosing any two distinct points on the line and finding the ratio of the change in y-values to the change in x-values. If the slope is positive, the line points up and to the right; if it is negative, the line points down and to the right (see Figure 9.24).

Spearman's r

It is inappropriate to calculate a **correlation coefficient** from paired data that are not strictly numerical. If the data are ordered in some fashion, it is possible to rank the data and to calculate the correlation coefficient from the ranks. In this case the calculation simplifies to the following formula:

$$r = 1 - \frac{6 \sum_{i=1}^{n} d_i^2}{n(n^2 - 1)}$$

Here n is the number of data points and d_i is the difference in values for point # i.

Standard Deviation

The standard deviation (σ) is a measure of how spread out a collection of numbers is. For a bell-shaped (i.e., normal) distribution, it measures the distance from the center of the curve to either inflection point (i.e., either point where the curve stops bending down and starts bending back up). With a normal distribution, there is a 68% probability of being within one standard deviation of the mean, and a 95% probability of being within two standard deviations of the mean.

Whether or not the curve is bell-shaped, the standard deviation is the square root of the **variance**.

Standard Error

The **standard deviation** $\sigma_{\bar{x}}$ of the sample mean $\bar{x}$ is known as the standard error, and is given by

$$\sigma_{\bar{x}} = \frac{\sigma}{\sqrt{n}}$$

The value of σ is the standard deviation of the original random variable and n is the **sample size**.

Standard Normal Random Variable

The standard normal random variable is a synonym for the **Z random variable**—a normal random variable with mean 0 and standard deviation 1.

Statistical Process Control

Statistical process control refers to the use of statistics to maintain quality in manufacturing, etc.

Stratified Sampling

In stratified sampling, the population is divided into groups that the experimenter wants to have represented; samples are then taken from each group and combined.

Summation Notation

Rather than always writing out long sums, if the terms all follow a given pattern it is more convenient to use summation notation. Summation notation consists of the upper-case Greek letter **sigma** ($\sum$), with an indexing letter and a range of values for the index, and with a formula for the terms. For example,

$$1^2 + 2^2 + 3^2 + 4^2 + 5^2 = \sum_{i=1}^{5} i^2$$

where i is the index, which runs from 1 to 5, and each term has the form i^2.

Symmetric

A graph of a probability distribution or a histogram is symmetric if it looks the same on either side of its center. For a symmetric distribution, the mean and median are both at the center.

For contrast, see **skewed to the right, skewed to the left**.

T Distribution

See **T random variable**.

T Random Variable

The T random variable is used in **hypothesis testing** and **confidence interval** calculation when small samples are chosen from populations that are roughly normal. It is formally defined as being proportional to the ratio of a **Z random variable** to the square root of a **chi-square random variable**. Consequently, it has degrees of freedom corresponding to the chi-square random variable appearing in its denominator.

In practice, only the critical values of a T random variable are needed; these are given in tables, indexed by the degrees of freedom (see page 209).

A T random variable can take any value, and the graph of its probability distribution is symmetric around 0. As the number of degrees of freedom increases, the T distribution approaches the standard normal distribution. The square of a T random variable is an F random variable.

The density function for m degrees of freedom is given by the formula:

$$\frac{\Gamma(m+1)/2}{\sqrt{m\pi}\,\Gamma(m/2)} \left(1 + \frac{x^2}{m}\right)^{-(m+1)/2}$$

See **gamma function**.

Figure 9.25 illustrates the density function for the T random variables with 2 and 60 degrees of freedom. Note that, as the degrees of freedom increases, there is more area near the peak of the distribution and less area in the tails.

T Statistic in Regression

For each coefficient β_i in a **regression** model, there is always the concern as to whether our estimate of it is meaningful, or if the coefficient itself is really 0 (i.e., the **independent variable** X_i doesn't contribute anything to the model). It is possible to calculate a t statistic for each coefficient to perform a **hypothesis test** on this question. The value of the t statistic is found by dividing the coefficient by the standard error for that coefficient. If the null hypothesis is true, and the coefficient truly has a value of zero, then the t statistic will come from a t distribution with $n - m$ degrees of freedom, where n is the number of observations and m is the number of parameters (which is one more than the number of independent variables). Look in the t distribution table (page 209) for the critical value with $n - m$ degrees of freedom. Reject the null hypothesis if the absolute value of the calculated t statistic is greater than the critical value from the table. See also **multiple regression, F statistic in regression**.

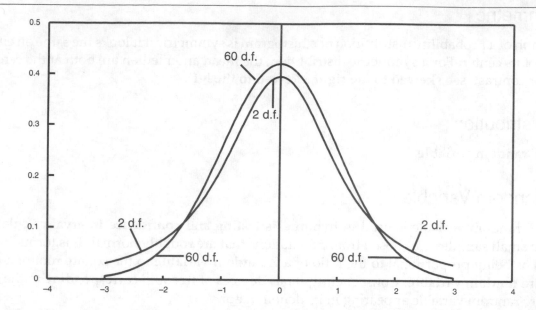

Figure 9.25: Comparison of the density function for two t distributions: 2 and 60 degrees of freedom

Tchebysheff's Theorem

Tchebysheff's Theorem establishes an upper bound for how many **standard deviations** a **random variable** can be from its mean. The theorem is very general, since no assumption is made about the probability distribution for the random variable. On the other hand, better upper bounds can usually be found if there is specific information about the random variable.

If Y is the random variable, with mean μ and standard deviation σ, then Tchebysheff's Theorem says that, for any positive constant k:

$$\Pr(|Y - \mu| \geq k\sigma) \leq \frac{1}{k^2}$$

For example, if $k = 2$:

$$\Pr(|Y - \mu| \geq 2\sigma) \leq \frac{1}{4}$$

Another way to say it is that for any random variable, there is at least a 75% chance that the value of the variable will be within two standard deviations of its mean. You may find a higher value for a specific distribution (for example, the normal distribution has about a 95% chance of being within two standard deviations of the mean), but you will never find any random variable distribution with probability less than 75% of being within two standard deviations of the mean.

Test Statistic

A test statistic is a **sample statistic** that is used in a **hypothesis test**. Common examples include the **chi-square random variable**, the **F random variable**, the **T random variable**, and the **Z random variable**. The test statistic must be designed so that if the null hypothesis is

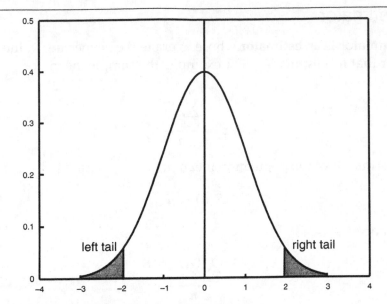

Figure 9.26: Critical region for two-tailed Z test. For the 5% significance level, there is 2.5% of the area in each tail. The critical values are -1.96 and 1.96.

true, the statistic will come from a known distribution. If the observed value of the test statistic is such that it is unlikely that it really did come from that known distribution, then there is evidence that the null hypothesis should be rejected.

Two-Tailed Hypothesis Test

A two-tailed hypothesis test is a **hypothesis test** where the null hypothesis claims that a population parameter is equal to a constant; in that case there are two ways that the null hypothesis can be wrong, two critical values, and two rejection regions (critical regions). For contrast, see **left-tailed hypothesis test, right-tailed hypothesis test**. Figure 9.26 illustrates the two critical values for a test using the Z (standard normal) distribution.

Type I Error

A Type I error consists of rejecting the null hypothesis in a **hypothesis test** when it is really true. The probability of this error occurring is equal to the significance level.

Type II Error

A Type II error consists of accepting the null hypothesis in a **hypothesis test** when it is false. Since there are many ways that the null hypothesis can be false, there is no single probability of this error occurring; rather, the probability is a function of the actual value of the population parameter being tested. See **power of test**.

Unbiased Estimator

An unbiased estimator is an **estimator** whose average (i.e., expected) value is exactly equal to the **parameter** that it is estimating. For example, the sample mean

$$\bar{x} = \frac{\sum\limits_{i=1}^{n} X_i}{n}$$

is an unbiased estimator of the population mean μ:

$$E(\bar{x}) = \mu$$

The sample variance

$$s^2 = \frac{\sum\limits_{i=1}^{n}(X_i - \bar{x})^2}{n-1}$$

is an unbiased estimator of the population variance σ^2:

$$E(s^2) = \sigma^2$$

Uniform Random Variable

If an event is equally likely to occur at any time within an interval $a \leq X \leq b$, then the time of its occurrence X is a uniform random variable. Its probability density function is given by:

$$f(x) = \begin{cases} \frac{1}{b-a} & a \leq x \leq b \\ 0 & \text{otherwise} \end{cases}$$

Its mean is $\mu = \frac{(a+b)}{2}$, and its variance is $\sigma^2 = \frac{(b-a)^2}{12}$. Figure 9.27 illustrates the density function for a uniform random variable.

Variance

The variance is a measure of the spread of a collection of numbers; for a population it is symbolized by σ^2 (sigma-squared). It is the average of the square of the differences between each number and the population **mean** μ:

$$\sigma^2 = \frac{(X_1 - \mu)^2 + (X_2 - \mu)^2 + \ldots + (X_n - \mu)^2}{n}$$

The variance can also be found from this formula:

$$\sigma^2 = \overline{x^2} - \bar{x}^2$$

For a sample the calculation is similar except that instead of dividing by the sample size, we divide by one less than the sample size. The sample variance is symbolized by s^2:

$$s^2 = \frac{(X_1 - \bar{x})^2 + (X_2 - \bar{x})^2 + \ldots + (X_n - \bar{x})^2}{n-1}$$

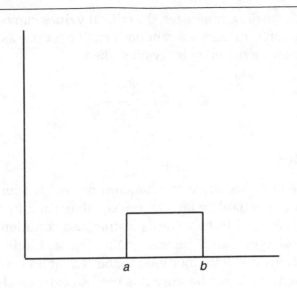

Figure 9.27: Probability density function for uniform random variable: height $= \frac{1}{b-a}$

The quantity $\bar{x}$ is the sample average.

The variance in either case is the square of the **standard deviation**.

The variance of a random variable expresses a similar concept—it is a measure of how spread out the probabilities are. The variance of a discrete random variable X (written $\text{Var}(X)$ or σ^2) can be found from the formula:

$$f(X_1)[X_1 - E(X)]^2 + f(X_2)[X_2 - E(X)]^2 + \ldots + f(X_m)[X_m - E(X)]^2$$

The X_1 to X_m are the m possible values of X, $E(X)$ is the expected value of X, and $f(X_i)$ is the probability that X takes on the value X_i. In other words, f is the probability function for X. The variance can also be found from this formula (which works for discrete or continuous random variables):

$$\text{Var}(X) = E(X^2) - [E(X)]^2$$

The variance of a random variable satisfies these properties:

- $\text{Var}(X)$ is always nonnegative; it is zero if and only if X is a constant.

- $\text{Var}(cX) = c^2 \text{Var}(X)$, where c is a constant.

- $\text{Var}(X + Y) = \text{Var}(X) + \text{Var}(Y)$, if X and Y are independent random variables.

- $\text{Var}(X) + \text{Var}(Y) = \text{Var}(X) + \text{Var}(Y) + 2\,\text{Cov}(X, Y)$, in general. The quantity $\text{Cov}(X, Y)$ is the **covariance** between X and Y.

Wilcoxon Signed Rank Test

The Wilcoxon signed rank test is a **nonparametric** hypothesis test similar to the **sign test**, but where the sign test ignores the magnitude of the differences between the pairs of observations, the Wilcoxon signed rank test ranks the differences by magnitude, and then compares the sums of the ranks that favor the first observation with the sum of the ranks that favor the second observation. For small sample sizes, the **critical values** can be read from a table. For larger sample sizes, we can construct a Z random variable for use as a **test statistic**: if T^+ is the sum of the ranks favoring the first observation, then

$$Z = \frac{T^+ - \frac{n(n+1)}{4}}{\sqrt{\frac{n(n+1)(2n+1)}{24}}}$$

Z Random Variable

The Z random variable (also known as the standard normal random variable) is a special case of the **normal random variable** with its mean equal to 0 and its variance (and standard deviation) equal to 1. It is used in **hypothesis testing** and **confidence interval** calculation where the sample size is larger than 30 because of the **Central Limit Theorem**.

All probability questions involving normal random variables can be converted to questions about the Z random variable. Because it is used so extensively, tables for Z random variables tend to be more extensive (see page 205 for a table).

APPENDIX A

Statistical Tables

STANDARD NORMAL (Z) TABLE

Two-Tailed Standard Normal (Z) Table: $p = \Pr(-a < Z < a)$

This table gives the probability p that a standard normal random variable Z (mean 0, standard deviation 1) will be between $-a$ and a.

a	p	a	p
0.100	0.0796	1.400	0.8384
0.200	0.1586	**1.439**	**0.8500**
0.300	0.2358	1.500	0.8664
0.400	0.3108	1.600	0.8904
0.500	0.3830	**1.645**	**0.9000**
0.600	0.4514	1.700	0.9108
0.700	0.5160	1.800	0.9282
0.800	0.5762	1.900	0.9426
0.900	0.6318	**1.960**	**0.9500**
1.000	0.6826	2.000	0.9544
1.100	0.7286	2.500	0.9876
1.200	0.7698	**2.576**	**0.9900**
1.282	**0.8000**	3.000	0.9974
1.300	0.8064	3.100	0.9980

The values in boldface are those commonly used for confidence intervals and hypothesis testing.

The table on the next two pages gives the probability p that a standard normal random variable Z will be less than the specified value a: $p = \Pr(Z < a)$. Also, it gives the area under the standard normal density function to the left of the specified value a.

Here is the connection between the two tables. If $p_1 = \Pr(-a < Z < a)$ (the value from the one-tailed table), and $p_2 = \Pr(Z < a)$ (the value from the two-tailed table), then $p_1 = 2p_2 - 1$.

ONE-TAILED STANDARD NORMAL (*Z*) RANDOM VARIABLE TABLE:

$\Pr(Z < a) = p$ (page 1)

a	p	a	p	a	p	a	p	a	p	a	p
-2.99	.0014	-2.49	.0064	-1.99	.0233	-1.49	.0681	-0.99	.1611	-0.49	.3121
-2.98	.0014	-2.48	.0066	-1.98	.0239	-1.48	.0694	-0.98	.1635	-0.48	.3156
-2.97	.0015	-2.47	.0068	-1.97	.0244	-1.47	.0708	-0.97	.1660	-0.47	.3192
-2.96	.0015	-2.46	.0069	-1.96	.0250	-1.46	.0721	-0.96	.1685	-0.46	.3228
-2.95	.0016	-2.45	.0071	-1.95	.0256	-1.45	.0735	-0.95	.1711	-0.45	.3264
-2.94	.0016	-2.44	.0073	-1.94	.0262	-1.44	.0749	-0.94	.1736	-0.44	.3300
-2.93	.0017	-2.43	.0075	-1.93	.0268	-1.43	.0764	-0.93	.1762	-0.43	.3336
-2.92	.0018	-2.42	.0078	-1.92	.0274	-1.42	.0778	-0.92	.1788	-0.42	.3372
-2.91	.0018	-2.41	.0080	-1.91	.0281	-1.41	.0793	-0.91	.1814	-0.41	.3409
-2.90	.0019	-2.40	.0082	-1.90	.0287	-1.40	.0808	-0.90	.1841	-0.40	.3446
-2.89	.0019	-2.39	.0084	-1.89	.0294	-1.39	.0823	-0.89	.1867	-0.39	.3483
-2.88	.0020	-2.38	.0087	-1.88	.0301	-1.38	.0838	-0.88	.1894	-0.38	.3520
-2.87	.0021	-2.37	.0089	-1.87	.0307	-1.37	.0853	-0.87	.1922	-0.37	.3557
-2.86	.0021	-2.36	.0091	-1.86	.0314	-1.36	.0869	-0.86	.1949	-0.36	.3594
-2.85	.0022	-2.35	.0094	-1.85	.0322	-1.35	.0885	-0.85	.1977	-0.35	.3632
-2.84	.0023	-2.34	.0096	-1.84	.0329	-1.34	.0901	-0.84	.2005	-0.34	.3669
-2.83	.0023	-2.33	.0099	-1.83	.0336	-1.33	.0918	-0.83	.2033	-0.33	.3707
-2.82	.0024	-2.32	.0102	-1.82	.0344	-1.32	.0934	-0.82	.2061	-0.32	.3745
-2.81	.0025	-2.31	.0104	-1.81	.0351	-1.31	.0951	-0.81	.2090	-0.31	.3783
-2.80	.0026	-2.30	.0107	-1.80	.0359	-1.30	.0968	-0.80	.2119	-0.30	.3821
-2.79	.0026	-2.29	.0110	-1.79	.0367	-1.29	.0985	-0.79	.2148	-0.29	.3859
-2.78	.0027	-2.28	.0113	-1.78	.0375	-1.28	.1003	-0.78	.2177	-0.28	.3897
-2.77	.0028	-2.27	.0116	-1.77	.0384	-1.27	.1020	-0.77	.2206	-0.27	.3936
-2.76	.0029	-2.26	.0119	-1.76	.0392	-1.26	.1038	-0.76	.2236	-0.26	.3974
-2.75	.0030	-2.25	.0122	-1.75	.0401	-1.25	.1056	-0.75	.2266	-0.25	.4013
-2.74	.0031	-2.24	.0125	-1.74	.0409	-1.24	.1075	-0.74	.2296	-0.24	.4052
-2.73	.0032	-2.23	.0129	-1.73	.0418	-1.23	.1093	-0.73	.2327	-0.23	.4090
-2.72	.0033	-2.22	.0132	-1.72	.0427	-1.22	.1112	-0.72	.2358	-0.22	.4129
-2.71	.0034	-2.21	.0136	-1.71	.0436	-1.21	.1131	-0.71	.2389	-0.21	.4168
-2.70	.0035	-2.20	.0139	-1.70	.0446	-1.20	.1151	-0.70	.2420	-0.20	.4207
-2.69	.0036	-2.19	.0143	-1.69	.0455	-1.19	.1170	-0.69	.2451	-0.19	.4247
-2.68	.0037	-2.18	.0146	-1.68	.0465	-1.18	.1190	-0.68	.2483	-0.18	.4286
-2.67	.0038	-2.17	.0150	-1.67	.0475	-1.17	.1210	-0.67	.2514	-0.17	.4325
-2.66	.0039	-2.16	.0154	-1.66	.0485	-1.16	.1230	-0.66	.2546	-0.16	.4364
-2.65	.0040	-2.15	.0158	-1.65	.0495	-1.15	.1251	-0.65	.2578	-0.15	.4404
-2.64	.0041	-2.14	.0162	-1.64	.0505	-1.14	.1271	-0.64	.2611	-0.14	.4443
-2.63	.0043	-2.13	.0166	-1.63	.0516	-1.13	.1292	-0.63	.2643	-0.13	.4483
-2.62	.0044	-2.12	.0170	-1.62	.0526	-1.12	.1314	-0.62	.2676	-0.12	.4522
-2.61	.0045	-2.11	.0174	-1.61	.0537	-1.11	.1335	-0.61	.2709	-0.11	.4562
-2.60	.0047	-2.10	.0179	-1.60	.0548	-1.10	.1357	-0.60	.2743	-0.10	.4602
-2.59	.0048	-2.09	.0183	-1.59	.0559	-1.09	.1379	-0.59	.2776	-0.09	.4641
-2.58	.0049	-2.08	.0188	-1.58	.0570	-1.08	.1401	-0.58	.2810	-0.08	.4681
-2.57	.0051	-2.07	.0192	-1.57	.0582	-1.07	.1423	-0.57	.2843	-0.07	.4721
-2.56	.0052	-2.06	.0197	-1.56	.0594	-1.06	.1446	-0.56	.2877	-0.06	.4761
-2.55	.0054	-2.05	.0202	-1.55	.0605	-1.05	.1469	-0.55	.2912	-0.05	.4801
-2.54	.0055	-2.04	.0207	-1.54	.0618	-1.04	.1492	-0.54	.2946	-0.04	.4840
-2.53	.0057	-2.03	.0212	-1.53	.0630	-1.03	.1515	-0.53	.2981	-0.03	.4880
-2.52	.0059	-2.02	.0217	-1.52	.0642	-1.02	.1539	-0.52	.3015	-0.02	.4920
-2.51	.0060	-2.01	.0222	-1.51	.0655	-1.01	.1562	-0.51	.3050	-0.01	.4960
-2.50	.0062	-2.00	.0228	-1.50	.0668	-1.00	.1587	-0.50	.3085	0.00	.5000

ONE-TAILED STANDARD NORMAL (Z) RANDOM VARIABLE TABLE:

$\Pr(Z < a) = p$ (page 2)

a	p	a	p	a	p	a	p	a	p	a	p
0.01	.5040	0.51	.6950	1.01	.8438	1.51	.9345	2.01	.9778	2.51	.9940
0.02	.5080	0.52	.6985	1.02	.8461	1.52	.9357	2.02	.9783	2.52	.9941
0.03	.5120	0.53	.7019	1.03	.8485	1.53	.9370	2.03	.9788	2.53	.9943
0.04	.5160	0.54	.7054	1.04	.8508	1.54	.9382	2.04	.9793	2.54	.9945
0.05	.5199	0.55	.7088	1.05	.8531	1.55	.9394	2.05	.9798	2.55	.9946
0.06	.5239	0.56	.7123	1.06	.8554	1.56	.9406	2.06	.9803	2.56	.9948
0.07	.5279	0.57	.7157	1.07	.8577	1.57	.9418	2.07	.9808	2.57	.9949
0.08	.5319	0.58	.7190	1.08	.8599	1.58	.9429	2.08	.9812	2.58	.9951
0.09	.5359	0.59	.7224	1.09	.8621	1.59	.9441	2.09	.9817	2.59	.9952
0.10	.5398	0.60	.7257	1.10	.8643	1.60	.9452	2.10	.9821	2.60	.9953
0.11	.5438	0.61	.7291	1.11	.8665	1.61	.9463	2.11	.9826	2.61	.9955
0.12	.5478	0.62	.7324	1.12	.8686	1.62	.9474	2.12	.9830	2.62	.9956
0.13	.5517	0.63	.7357	1.13	.8708	1.63	.9484	2.13	.9834	2.63	.9957
0.14	.5557	0.64	.7389	1.14	.8729	1.64	.9495	2.14	.9838	2.64	.9959
0.15	.5596	0.65	.7422	1.15	.8749	1.65	.9505	2.15	.9842	2.65	.9960
0.16	.5636	0.66	.7454	1.16	.8770	1.66	.9515	2.16	.9846	2.66	.9961
0.17	.5675	0.67	.7486	1.17	.8790	1.67	.9525	2.17	.9850	2.67	.9962
0.18	.5714	0.68	.7517	1.18	.8810	1.68	.9535	2.18	.9854	2.68	.9963
0.19	.5753	0.69	.7549	1.19	.8830	1.69	.9545	2.19	.9857	2.69	.9964
0.20	.5793	0.70	.7580	1.20	.8849	1.70	.9554	2.20	.9861	2.70	.9965
0.21	.5832	0.71	.7611	1.21	.8869	1.71	.9564	2.21	.9864	2.71	.9966
0.22	.5871	0.72	.7642	1.22	.8888	1.72	.9573	2.22	.9868	2.72	.9967
0.23	.5910	0.73	.7673	1.23	.8907	1.73	.9582	2.23	.9871	2.73	.9968
0.24	.5948	0.74	.7704	1.24	.8925	1.74	.9591	2.24	.9875	2.74	.9969
0.25	.5987	0.75	.7734	1.25	.8944	1.75	.9599	2.25	.9878	2.75	.9970
0.26	.6026	0.76	.7764	1.26	.8962	1.76	.9608	2.26	.9881	2.76	.9971
0.27	.6064	0.77	.7794	1.27	.8980	1.77	.9616	2.27	.9884	2.77	.9972
0.28	.6103	0.78	.7823	1.28	.8997	1.78	.9625	2.28	.9887	2.78	.9973
0.29	.6141	0.79	.7852	1.29	.9015	1.79	.9633	2.29	.9890	2.79	.9974
0.30	.6179	0.80	.7881	1.30	.9032	1.80	.9641	2.30	.9893	2.80	.9974
0.31	.6217	0.81	.7910	1.31	.9049	1.81	.9649	2.31	.9896	2.81	.9975
0.32	.6255	0.82	.7939	1.32	.9066	1.82	.9656	2.32	.9898	2.82	.9976
0.33	.6293	0.83	.7967	1.33	.9082	1.83	.9664	2.33	.9901	2.83	.9977
0.34	.6331	0.84	.7995	1.34	.9099	1.84	.9671	2.34	.9904	2.84	.9977
0.35	.6368	0.85	.8023	1.35	.9115	1.85	.9678	2.35	.9906	2.85	.9978
0.36	.6406	0.86	.8051	1.36	.9131	1.86	.9686	2.36	.9909	2.86	.9979
0.37	.6443	0.87	.8078	1.37	.9147	1.87	.9693	2.37	.9911	2.87	.9979
0.38	.6480	0.88	.8106	1.38	.9162	1.88	.9699	2.38	.9913	2.88	.9980
0.39	.6517	0.89	.8133	1.39	.9177	1.89	.9706	2.39	.9916	2.89	.9981
0.40	.6554	0.90	.8159	1.40	.9192	1.90	.9713	2.40	.9918	2.90	.9981
0.41	.6591	0.91	.8186	1.41	.9207	1.91	.9719	2.41	.9920	2.91	.9982
0.42	.6628	0.92	.8212	1.42	.9222	1.92	.9726	2.42	.9922	2.92	.9982
0.43	.6664	0.93	.8238	1.43	.9236	1.93	.9732	2.43	.9925	2.93	.9983
0.44	.6700	0.94	.8264	1.44	.9251	1.94	.9738	2.44	.9927	2.94	.9984
0.45	.6736	0.95	.8289	1.45	.9265	1.95	.9744	2.45	.9929	2.95	.9984
0.46	.6772	0.96	.8315	1.46	.9279	1.96	.9750	2.46	.9931	2.96	.9985
0.47	.6808	0.97	.8340	1.47	.9292	1.97	.9756	2.47	.9932	2.97	.9985
0.48	.6844	0.98	.8365	1.48	.9306	1.98	.9761	2.48	.9934	2.98	.9986
0.49	.6879	0.99	.8389	1.49	.9319	1.99	.9767	2.49	.9936	2.99	.9986
0.50	.6915	1.00	.8413	1.50	.9332	2.00	.9772	2.50	.9938	3.00	.9987

CHI SQUARE TABLE

The table gives the value of a such that $\Pr(\chi^2_{DF} < a) = p$, where χ^2_{DF} is a chi-square random variable with DF degrees of freedom. For example, there is a probability of .95 that a chi-square random variable with 6 degrees of freedom will be less than 12.6.

DF	p = .005	.01	.025	.05	.25	.5	.75	.9	.95	.975	.99
1	.000	.000	.001	.004	.10	.45	1.32	2.71	3.84	5.02	6.64
2	.010	.020	.051	.10	.58	1.39	2.77	4.61	5.99	7.38	9.21
3	.072	.11	.22	.35	1.21	2.37	4.11	6.25	7.81	9.35	11.3
4	.21	.30	.48	.71	1.92	3.36	5.39	7.78	9.49	11.1	13.3
5	.41	.55	.83	1.15	2.67	4.35	6.63	9.24	11.1	12.8	15.1
6	.68	.87	1.24	1.64	3.45	5.35	7.84	10.6	12.6	14.4	16.8
7	.99	1.24	1.69	2.17	4.25	6.35	9.04	12.0	14.1	16.0	18.5
8	1.34	1.65	2.18	2.73	5.07	7.34	10.2	13.4	15.5	17.5	20.1
9	1.73	2.09	2.70	3.33	5.90	8.34	11.4	14.7	16.9	19.0	21.7
10	2.16	2.56	3.25	3.94	6.74	9.34	12.5	16.0	18.3	20.5	23.2
11	2.60	3.05	3.82	4.57	7.58	10.3	13.7	17.3	19.7	21.9	24.7
12	3.07	3.57	4.40	5.23	8.44	11.3	14.8	18.5	21.0	23.3	26.2
13	3.56	4.11	5.01	5.89	9.30	12.3	16.0	19.8	22.4	24.7	27.7
14	4.07	4.66	5.63	6.57	10.2	13.3	17.1	21.1	23.7	26.1	29.1
15	4.60	5.23	6.26	7.26	11.0	14.3	18.2	22.3	25.0	27.5	30.6
16	5.14	5.81	6.91	7.96	11.9	15.3	19.4	23.5	26.3	28.8	32.0
17	5.70	6.41	7.56	8.67	12.8	16.3	20.5	24.8	27.6	30.2	33.4
18	6.26	7.01	8.23	9.39	13.7	17.3	21.6	26.0	28.9	31.5	34.8
19	6.84	7.63	8.91	10.1	14.6	18.3	22.7	27.2	30.1	32.9	36.2
20	7.43	8.26	9.59	10.9	15.5	19.3	23.8	28.4	31.4	34.2	37.6
21	8.03	8.90	10.3	11.6	16.3	20.3	24.9	29.6	32.7	35.5	38.9
22	8.64	9.54	11.0	12.3	17.2	21.3	26.0	30.8	33.9	36.8	40.3
23	9.26	10.2	11.7	13.1	18.1	22.3	27.1	32.0	35.2	38.1	41.6
24	9.89	10.9	12.4	13.8	19.0	23.3	28.2	33.2	36.4	39.4	43.0
25	10.5	11.5	13.1	14.6	19.9	24.3	29.3	34.4	37.7	40.6	44.3
26	11.2	12.2	13.8	15.4	20.8	25.3	30.4	35.6	38.9	41.9	45.6
27	11.8	12.9	14.6	16.1	21.7	26.3	31.5	36.7	40.1	43.2	47.0
28	12.5	13.6	15.3	16.9	22.7	27.3	32.6	37.9	41.3	44.5	48.3
29	13.1	14.3	16.0	17.7	23.6	28.3	33.7	39.1	42.6	45.7	49.6
30	13.8	14.9	16.8	18.5	24.5	29.3	34.8	40.3	43.8	47.0	50.9
35	17.2	18.5	20.6	22.5	29.0	34.3	40.2	46.1	49.8	53.2	57.3
40	20.7	22.2	24.4	26.5	33.7	39.3	45.6	51.8	55.8	59.3	63.7
50	28.0	29.7	32.4	34.8	42.9	49.3	56.3	63.2	67.5	71.4	76.2
60	35.5	37.5	40.5	43.2	52.3	59.3	67.0	74.4	79.1	83.3	88.4
70	43.3	45.4	48.8	51.7	61.7	69.3	77.6	85.5	90.5	95.0	100.4
80	51.2	53.5	57.2	60.4	71.1	79.3	88.1	96.6	101.9	106.6	112.3
90	59.2	61.8	65.6	69.1	80.6	89.3	98.6	107.6	113.1	118.1	124.1
100	67.3	70.1	74.2	77.9	90.1	99.3	109.1	118.5	124.3	129.6	135.8

TWO-TAILED *T* DISTRIBUTION TABLE

If T is a random variable with a t distribution with DF degrees of freedom, then the table gives the value of a such that $\Pr(-a < T < a) = p$. For example, there is a 95% probability that a T random variable with 7 degrees of freedom will be between -2.365 and 2.365.

DF	p=.900	p=.950	p=.99
1	6.314	12.706	63.657
2	2.920	4.303	9.925
3	2.353	3.182	5.841
4	2.132	2.776	4.604
5	2.015	2.571	4.032
6	1.943	2.447	3.707
7	1.895	2.365	3.499
8	1.860	2.306	3.355
9	1.833	2.262	3.250
10	1.812	2.228	3.169
11	1.796	2.201	3.106
12	1.782	2.179	3.055
13	1.771	2.160	3.012
14	1.761	2.145	2.977
15	1.753	2.131	2.947
16	1.746	2.120	2.921
17	1.740	2.110	2.898
18	1.734	2.101	2.878
19	1.729	2.093	2.861
20	1.725	2.086	2.845
21	1.721	2.080	2.831
22	1.717	2.074	2.819
23	1.714	2.069	2.807
24	1.711	2.064	2.797
25	1.708	2.060	2.787
26	1.706	2.056	2.779
27	1.703	2.052	2.771
28	1.701	2.048	2.763
29	1.699	2.045	2.756
30	1.697	2.042	2.750
35	1.690	2.030	2.724
40	1.684	2.021	2.704
45	1.679	2.014	2.690
50	1.676	2.009	2.678
55	1.673	2.004	2.668
60	1.671	2.000	2.660
80	1.664	1.990	2.639
100	1.660	1.984	2.626
120	1.658	1.980	2.617

ONE-TAILED *T* DISTRIBUTION TABLE

If T is a random variable with a t distribution with DF degrees of freedom, then the table gives the value of a such that $\Pr(T < a) = p$. For example, there is a .975 probability that a T random variable with 7 degrees of freedom will be less than 2.365.

DF	$p=.750$	$p=.900$	$p=.950$	$p=.975$	$p=.990$	$p=.995$
1	1.000	3.078	6.314	12.706	31.821	63.657
2	0.816	1.886	2.920	4.303	6.965	9.925
3	0.765	1.638	2.353	3.182	4.541	5.841
4	0.741	1.533	2.132	2.776	3.747	4.604
5	0.727	1.476	2.015	2.571	3.365	4.032
6	0.718	1.440	1.943	2.447	3.143	3.707
7	0.711	1.415	1.895	2.365	2.998	3.499
8	0.706	1.397	1.860	2.306	2.896	3.355
9	0.703	1.383	1.833	2.262	2.821	3.250
10	0.700	1.372	1.812	2.228	2.764	3.169
11	0.697	1.363	1.796	2.201	2.718	3.106
12	0.695	1.356	1.782	2.179	2.681	3.055
13	0.694	1.350	1.771	2.160	2.650	3.012
14	0.692	1.345	1.761	2.145	2.624	2.977
15	0.691	1.341	1.753	2.131	2.602	2.947
16	0.690	1.337	1.746	2.120	2.583	2.921
17	0.689	1.333	1.740	2.110	2.567	2.898
18	0.688	1.330	1.734	2.101	2.552	2.878
19	0.688	1.328	1.729	2.093	2.539	2.861
20	0.687	1.325	1.725	2.086	2.528	2.845
21	0.686	1.323	1.721	2.080	2.518	2.831
22	0.686	1.321	1.717	2.074	2.508	2.819
23	0.685	1.319	1.714	2.069	2.500	2.807
24	0.685	1.318	1.711	2.064	2.492	2.797
25	0.684	1.316	1.708	2.060	2.485	2.787
26	0.684	1.315	1.706	2.056	2.479	2.779
27	0.684	1.314	1.703	2.052	2.473	2.771
28	0.683	1.313	1.701	2.048	2.467	2.763
29	0.683	1.311	1.699	2.045	2.462	2.756
30	0.683	1.310	1.697	2.042	2.457	2.750
35	0.682	1.306	1.690	2.030	2.438	2.724
40	0.681	1.303	1.684	2.021	2.423	2.704
45	0.680	1.301	1.679	2.014	2.412	2.690
50	0.679	1.299	1.676	2.009	2.403	2.678
55	0.679	1.297	1.673	2.004	2.396	2.668
60	0.679	1.296	1.671	2.000	2.390	2.660
80	0.678	1.292	1.664	1.990	2.374	2.639
100	0.677	1.290	1.660	1.984	2.364	2.626
120	0.677	1.289	1.658	1.980	2.358	2.617

F DISTRIBUTION TABLE

In each of the following three tables, the numerator degrees of freedom are read along the top, and the denominator degrees of freedom are read along the left side. The table gives the value of a such that $\Pr(F < a) = p$, where F is a random variable with an F distribution with DF_{num} numerator degrees of freedom and DF_{den} denominator degrees of freedom. Each table has a different value of p; first .99; then .95; then .90. These are values commonly used for hypothesis testing. Another way to describe the table is to give the value of $1 - p$, which is the right tail area (that is, the area to the right of the given value of a). For example, there is a .99 probability that an F random variable with 5 numerator degrees of freedom and 9 denominator degrees of freedom will be less than 6.06.

$\Pr(F < a) = .99$; right tail area $= .01$

DF_{den}	$DF_{num} =$	2	3	4	5	10	15	20	30	60	120
2		99.00	99.16	99.25	99.30	99.40	99.43	99.45	99.47	99.48	99.49
3		30.82	29.46	28.71	28.24	27.23	26.87	26.69	26.50	26.32	26.22
4		18.00	16.69	15.98	15.52	14.55	14.20	14.02	13.84	13.65	13.56
5		13.27	12.06	11.39	10.97	10.05	9.72	9.55	9.38	9.20	9.11
6		10.92	9.78	9.15	8.75	7.87	7.56	7.40	7.23	7.06	6.97
7		9.55	8.45	7.85	7.46	6.62	6.31	6.16	5.99	5.82	5.74
8		8.65	7.59	7.01	6.63	5.81	5.52	5.36	5.20	5.03	4.95
9		8.02	6.99	6.42	6.06	5.26	4.96	4.81	4.65	4.48	4.40
10		7.56	6.55	5.99	5.64	4.85	4.56	4.41	4.25	4.08	4.00
15		6.36	5.42	4.89	4.56	3.80	3.52	3.37	3.21	3.05	2.96
20		5.85	4.94	4.43	4.10	3.37	3.09	2.94	2.78	2.61	2.52
30		5.39	4.51	4.02	3.70	2.98	2.70	2.55	2.39	2.21	2.11
60		4.98	4.13	3.65	3.34	2.63	2.35	2.20	2.03	1.84	1.73
120		4.79	3.95	3.48	3.17	2.47	2.19	2.03	1.86	1.66	1.53

$\Pr(F < a) = .95$; right tail area $= .05$

DF_{den}	$DF_{num} =$	2	3	4	5	10	15	20	30	60	120
2		19.00	19.16	19.25	19.30	19.40	19.43	19.45	19.46	19.48	19.49
3		9.55	9.28	9.12	9.01	8.79	8.70	8.66	8.62	8.57	8.55
4		6.94	6.59	6.39	6.26	5.96	5.86	5.80	5.75	5.69	5.66
5		5.79	5.41	5.19	5.05	4.74	4.62	4.56	4.50	4.43	4.40
6		5.14	4.76	4.53	4.39	4.06	3.94	3.87	3.81	3.74	3.70
7		4.74	4.35	4.12	3.97	3.64	3.51	3.44	3.38	3.30	3.27
8		4.46	4.07	3.84	3.69	3.35	3.22	3.15	3.08	3.01	2.97
9		4.26	3.86	3.63	3.48	3.14	3.01	2.94	2.86	2.79	2.75
10		4.10	3.71	3.48	3.33	2.98	2.85	2.77	2.70	2.62	2.58
15		3.68	3.29	3.06	2.90	2.54	2.40	2.33	2.25	2.16	2.11
20		3.49	3.10	2.87	2.71	2.35	2.20	2.12	2.04	1.95	1.90
30		3.32	2.92	2.69	2.53	2.16	2.01	1.93	1.84	1.74	1.68
60		3.15	2.76	2.53	2.37	1.99	1.84	1.75	1.65	1.53	1.47
120		3.07	2.68	2.45	2.29	1.91	1.75	1.66	1.55	1.43	1.35

$\Pr(F < a) = .90$; right tail area $= .10$

DF_{den}	$DF_{num} =$	2	3	4	5	10	15	20	30	60	120
2		9.00	9.16	9.24	9.29	9.39	9.42	9.44	9.46	9.47	9.48
3		5.46	5.39	5.34	5.31	5.23	5.20	5.18	5.17	5.15	5.14
4		4.32	4.19	4.11	4.05	3.92	3.87	3.84	3.82	3.79	3.78
5		3.78	3.62	3.52	3.45	3.30	3.24	3.21	3.17	3.14	3.12
6		3.46	3.29	3.18	3.11	2.94	2.87	2.84	2.80	2.76	2.74
7		3.26	3.07	2.96	2.88	2.70	2.63	2.59	2.56	2.51	2.49
8		3.11	2.92	2.81	2.73	2.54	2.46	2.42	2.38	2.34	2.32
9		3.01	2.81	2.69	2.61	2.42	2.34	2.30	2.25	2.21	2.18
10		2.92	2.73	2.61	2.52	2.32	2.24	2.20	2.16	2.11	2.08
15		2.70	2.49	2.36	2.27	2.06	1.97	1.92	1.87	1.82	1.79
20		2.59	2.38	2.25	2.16	1.94	1.84	1.79	1.74	1.68	1.64
30		2.49	2.28	2.14	2.05	1.82	1.72	1.67	1.61	1.54	1.50
60		2.39	2.18	2.04	1.95	1.71	1.60	1.54	1.48	1.40	1.35
120		2.35	2.13	1.99	1.90	1.65	1.55	1.48	1.41	1.32	1.26

APPENDIX B

Calculations

STATISTICAL CALCULATIONS ON A CALCULATOR

Important statistical calculations usually involve large amounts of data, more than anyone would want to try to analyze by hand. Fortunately calculators and computers are designed to handle repetitive arithmetic calculations. Most scientific calculators available today have a wide range of statistical features. (Software packages for analyzing statistics will be discussed in the next section.)

What are the advantages and disadvantages of using a calculator instead of a computer software package? Calculators are portable and affordable, but there are limits to the amount of data that they can handle.

Since there are many different types of calculators on the market, we will not list the keystrokes for every single one; rather we will try to summarize the common procedures necessary to perform statistical analyses on a calculator. Once you purchase one yourself, it will be easier then to consult the owner's manual for your calculator.

1. Entering data. With improvements in the way calculators display data, it has become much easier to enter and edit data. Typically you will need to choose a STATISTICS menu on your calculator and then choose either a DATA or EDIT sub-menu. By using the arrow keys and where appropriate the DELETE key, you can enter your data into one or more lists or columns for analysis. Again, using the arrow keys, you can proofread and if necessary correct your data. (Remember this basic principle of statistical analysis: garbage in, garbage out.)

2. Saving data. One of the basic principles in working with calculators and computers is that you should never have to type in the same set of numbers twice. If there is any chance that you will want to analyze your data set further at a later date, it would be a good idea to save your data. Some calculators will provide you with a limited choice

of names for your data set, while others will allow you to choose any name. You can then recall the data set at a later date when you need it.

3. Transferring data. If you are working with someone else who has the same kind of calculator, you may wish to share your data with them. This is possible with many scientific calculators, either through a cable that is plugged in to both calculators or via an infrared data link.

4. Statistical calculations. Going back to the STATISTICS menu on your calculator, you will usually find the following options under a CALC sub-menu:

   - basic descriptive statistics (mean, variance, standard deviation, minimum, maximum)
   - various types of regression models
   - hypothesis tests on some later model calculators
   - statistical displays. If there is a STATISTICS PLOT option on your calculator, you will be able to graph frequency histograms, scatterplots (and simple regression models superimposed over the data points), and box-plots.

STATISTICAL CALCULATIONS USING A COMPUTER SOFTWARE PACKAGE

For most practical applications of statistics, it is best to use a computer software package such as SAS, Minitab, SPSS, etc. Why?

- A computer software package can handle much more in the way of arithmetic calculations than you can by hand or with a calculator.

- A computer software package can offer a much wider array of statistical calculations and types of data displays than a calculator can.

- It is possible to insert the results of your work directly into a word-processing document.

- It is easier to transmit your data and/or analysis to someone else, either by floppy disk or via electronic mail.

- Computer displays can show more information in their screens at a given time than a calculator, making data review and display much easier.

Statistical software packages are sold for a variety of operating systems (DOS, Macintosh, Windows, UNIX, VMS, etc.) but all follow the same sort of basic operations:

1. Entering data. Data can be entered at the keyboard (which can involve a substantial amount of work for large data sets) or loaded from an existing data file. Data can be entered or stored outside of the statistical software package in an ASCII file, but it is usually easier to enter the data from within the package itself. Typically the data will be displayed in a spreadsheet format; you may then use the arrow keys to move around to review or edit data.

2. Saving data. One of the basic principles in working with calculators and computers is that you should never have to type in the same set of numbers twice. If there is any chance that you will want to analyze your data set further at a later date, it would be a good idea to save your data. Software packages will always prompt you before you exit them to save your data first into a data file; even so, it is best to save your data as soon as you have entered it in case of a system crash.

3. Transferring data and results to a word-processing document. Within the Windows and Macintosh operating systems it is possible to save your data and/or results onto a clip board and then insert them into your document. You can also dynamically link the document to your data file so that if you change your data file at a later date the document will automatically be updated.

4. Statistical calculations. Statistical software packages will perform for you every calculation described in this book. The functions will typically be grouped together into menus. E.g., descriptive statistics such as the mean, variance, standard deviation, etc. will typically be contained in a menu item "One-variable statistics," regression models in a menu item "Regression," etc.

5. Graphs. Statistical software packages will create for you every display discussed in this book. The different kinds of graphs will typically be contained in a menu item "Plot."

If you are working from a DOS, VMS, or UNIX operating system you will start your program by typing a command at the command prompt. If you are working from a Windows or Macintosh environment you will start your program by clicking with your mouse on the program's icon. Once the program is running, you should be able to get specific advice on any topic from the program's "Help" feature (usually listed as a menu item).

STATISTICAL FUNCTIONS IN THE MICROSOFT EXCEL SPREADSHEET

A third possible choice of tools for statistical calculations is a general purpose spreadsheet, such as Microsoft Excel. If you use a spreadsheet for other purposes anyway, then learning the spreadsheet commands would be a good way to do statistical calculations.

The following lists some examples of the statistical functions included with newer versions of Microsoft Excel.

Description of Function	Example	Result
BINOMDIST(k, n, p,FALSE) calculates the probability that $X = k$, where X has a binomial distribution with parameters n and p.	=BINOMDIST(6,10,0.75,FALSE)	0.14600
BINOMDIST(k, n, p,TRUE) calculates the cumulative probability that X is less than or equal to k; in other words, it sums the probabilities from $X = 0$ up to $X = k$.	=BINOMDIST(6,10,0.75,TRUE)	0.22412
CHIDIST(a, df) gives the probability that a chi square random variable with df degrees of freedom will be greater than a	=CHIDIST(11.07,5)	0.05001
CHIINV(p, df) gives the value a such that $\Pr(\chi^2_{df} > a) = p$, where χ^2_{df} is a chi square random variable with df degrees of freedom. This function is the inverse of the previous one.	=CHIINV(0.05,5)	11.07048
COMBIN(n, j) gives the number of combinations when j objects are selected from n objects.	=COMBIN(52,5)	2,598,960
FACT(n) gives $n!$ (n factorial)	=FACT(9)	362,880
FDIST(a, df_{num}, df_{den}) gives the probability that an F random variable with df_{num} and df_{den} degrees of freedom will be greater than a.	=FDIST(5.96,10,4)	0.05006
FINV(p, df_{num}, df_{den}) gives the value a such that $\Pr(F > a) = p$, where F is an F random variable with df_{num} and df_{den} degrees of freedom. This function is the inverse of the previous one.	=FINV(0.05,10,4)	5.96435

Description of Function	Example	Result
HYPGEOMDIST(k, ns, M, N) gives $\Pr(X = k)$, where X has a hypergeometric distribution with population size N, sample size ns, and M objects in the population of type M, and k objects in the sample of type M.	=HYPGEOMDIST(3,10,13,52)	0.27806
NORMDIST($x, mu, sigma$,TRUE) gives the probability that a normal random variable (with mean mu and standard deviation $sigma$) will be less than x.	=NORMDIST(12,10,2,TRUE)	0.84134
NORMINV($p, mu, sigma$) gives the value a such that $\Pr(X < a) = p$ where X has a normal distribution. This function is the inverse of the previous one.	=NORMINV(0.841345,10,2)	12.00000
NORMSDIST(a) gives $\Pr(Z < a)$, where Z has a standard normal distribution.	=NORMSDIST(1.96)	0.97500
NORMSINV(p) gives the value a such that $\Pr(Z < a) = p$), where Z has a standard normal distribution. This function is the inverse of the previous one.	=NORMSINV(0.975)	1.95996
TDIST($a, df, 1$) gives the one-tail probability for a t distribution: $\Pr(T > a)$, where T is a random variable with the T distribution with df degrees of freedom.	=TDIST(2.365,7,1)	0.02499
TDIST($a, df, 2$) gives the two-tail probability for a t distribution: $\Pr(T > a) + \Pr(T < -a)$, or $2\Pr(T > a)$.	=TDIST(2.365,7,2)	0.04997
TINV(p, df) gives the value a such that $\Pr(T > a) = p$, where T is a random variable with the T distribution with df degrees of freedom.	=TINV(0.05,7)	2.36462

Excel also provides several functions that calculate descriptive statistics for a range of data:

AVERAGE(*range*)   calculate average
STDEVP(*range*)   standard deviation of a population
STDEV(*range*)   standard deviation of a sample
MEDIAN(*range*)   calculate median
PERCENTRANK(*range,value*)   give the percentile rank of *value* within the list *range*

These functions can be used for simple regression and correlation:

SLOPE(*yrange, xrange*)   slope of simple regression line
INTERCEPT(*yrange, xrange*) y intercept of simple regression line
RSQ(*yrange, xrange*) r squared value of simple regression
CORREL(*yrange, xrange*)   correlation coefficient between two ranges
See the HELP menu for more details on these and other statistical functions.

Index